Nanomaterials in Liquid Crystals

Nanomaterials in Liquid Crystals

Special Issue Editor
Ingo Dierking

MDPI • Basel • Beijing • Wuhan • Barcelona • Belgrade

MDPI

Special Issue Editor
Ingo Dierking
University of Manchester
UK

Editorial Office
MDPI
St. Alban-Anlage 66
Basel, Switzerland

This is a reprint of articles from the Special Issue published online in the open access journal *Nanomaterials* (ISSN 2079-4991) from 2017 to 2018 (available at: http://www.mdpi.com/journal/nanomaterials/special_issues/nano_liquid_crys)

For citation purposes, cite each article independently as indicated on the article page online and as indicated below:

LastName, A.A.; LastName, B.B.; LastName, C.C. Article Title. *Journal Name* **Year**, *Article Number*, Page Range.

ISBN 978-3-03897-115-3 (Pbk)
ISBN 978-3-03897-116-0 (PDF)

Cover image courtesy of Shakhawan Al-Zangana.

Contents

About the Special Issue Editor

Ingo Dierking received his Diploma in Physics in 1992 and his PhD at the Institute of Physical Chemistry of Clausthal University of Technology in 1995, after which he joined the IBM T.J. Watson Research Centre in Yorktown Heights, USA, as a postdoc. In 1997, he was awarded a Feodor-Lynen Fellowship from the Humboldt Foundation to work at the Physics Department of Chalmers University of Technology, Gothenburg, Sweden, being appointed as Docent for Physics in 1999. He then joined the Department of Physical Chemistry at the University of Darmstadt, where he remained until 2002 as a lecturer and received his Habilitation for Physical Chemistry, before moving to his present post at the School of Physics and Astronomy of the University of Manchester. Dierking is the 2009 recipient of the Hilsum Medal and the 2016 recipient of the Samsung Mid-Career Award for Research Excellence.

Preface to "Nanomaterials in Liquid Crystals"

During the recent years, research and development have led to liquid crystal materials being increasingly transformed from mesogenic mixtures to more complex systems of liquid crystal-based composites and nanoparticle dispersions. The mechanisms involved are threefold: (i) nanoparticles are added to liquid crystals in order to tune common liquid crystal material parameters, such as threshold voltage or response time, (ii) nanoparticles are dispersed to add and influence the final functionality and properties, such as ferromagnetism, ferroelectricity, semiconducting properties, fluorescence, or plasmonics, and (iii) anisotropic nanoparticles in an isotropic host can lead to the formation of lyotropic liquid crystals and, thus, to nanoparticle self-organization and self-assembly which are useful in nanotechnology and, particularly, in photonics and optoelectronics.

The present book results from a Special Issue of the journal "Nanomaterials" and covers the topic of nanomaterials in liquid crystals in a broad sense. With the topic being of rather multidisciplinary nature, aspects of nanoparticle synthesis, materials formulation, physical properties, applications, as well as theoretical descriptions are collected. The book content is roughly arranged into two parts, following the motivations of the research in colloidal liquid crystals over the recent years, I: Property Tuning and Added Functionality; II: Nanoparticle Self-Organization. Each part is introduced by a related review article, followed by reports of original research results. Part I starts with an introduction to "Ferroelectric Nanoparticles in Liquid Crystals", by Yuriy Garbovskiy and Anatoliy Glushchenko. The following articles then cover the synthesis of distinct iron oxide nanoparticles, the use of such particles to improve image sticking, and a description of the kinetics of ion capturing by nanomaterials. This section is rounded off by a discussion about the dynamic response of graphitic flakes in liquid crystals.

The opening review of Part II, by Ingo Dierking and Shakhawan Al-Zangana, provides a general overview of "Lyotropic Liquid Crystal Phases from Anisotropic Nanomaterials", discussing systems as diverse as inorganic and mineral liquid crystals, clays, biological nanoparticles from DNA to tobacco mosaic viruses, and cellulose nanocrystals, all the way to nanotubes, nanorods, and graphene oxide. The following articles cover phase-transition-driven nanoparticle assembly, thermally switchable self-assembled patterns of nanoparticles, magnetic nanoparticle-assisted optical patterns from spherical cholesterics, and polymer templating for tunable lasing.

The breadth of the contributions illustrates the fascinating possibilities of fundamental studies and applications that are offered by combining self-organization, anisotropy, liquid crystallinity, and nanomaterials. It is anticipated that this multidisciplinary field of research and development will lead to a wealth of novel systems in soft condensed matter, promising new applications in the areas of displays, optical devices and elements, meta-materials, sensors, drug delivery, and the like.

Ingo Dierking
Special Issue Editor

nanomaterials

MDPI

Editorial

Nanomaterials in Liquid Crystals

Ingo Dierking

School of Physics and Astronomy, University of Manchester, Oxford Road, Manchester M13 9PL, UK;
ingo.dierking@manchester.ac.uk

Received: 12 June 2018; Accepted: 14 June 2018; Published: 21 June 2018

Liquid crystals are often identified with the development of the flat panel television and computer screens that we all use on a daily basis. Despite their enormous success in this area, liquid crystal research is by far not exhausted and has reinvented itself, spearheading into other fields of research, due to their properties of self-organization, their fascinating optic and electro-optic properties, and their easy deformability and reorientation via electric, magnetic, mechanical and other external fields. Novel effects are being discovered, new modern and self-organized materials are constantly being developed and a whole range of non-display applications is being proposed, which are on the borderline between nanotechnology and soft condensed matter. Liquid crystals are also being employed as a vehicle to study fundamental physical questions, and proceeding into the areas of biology, nature and life. In this Special Issue of *Nanomaterials*, illustrative examples are introduced, which draw on aspects of self-organization of liquid crystals, colloidal ordering of nanoparticles, and the formation of anisotropic, liquid crystalline phases from nanoparticles. An exhaustive treatment of these topics up to about 2015 can be found in the two volume handbook edited by Lagerwall and Scalia [1].

Liquid crystals [2–4] are partially ordered, anisotropic fluids, which are thermodynamically located between the three dimensional solid crystal and the flow governed liquid. They exhibit orientational or low dimensional positional order of their long molecular axis or the molecular centres of mass, respectively, which results in anisotropic physical properties, such as refractive index, viscosity, elastic constant, electric conductivity, or magnetic susceptibility, while retaining the ability to flow. Liquid crystals are part of the ever growing and increasingly important family of soft condensed matter materials [5–9]. Two general classes of liquid crystals are mostly distinguished, *thermotropic* materials [10,11], which exhibit the liquid crystalline state exclusively on temperature variation, and *lyotropic* liquid crystals [12,13], where the formation of liquid crystal phases is achieved by concentration variation of shape anisotropic dopant materials in an isotropic carrier or host fluid. The latter type is most often composed from amphiphilic molecules in water, but can also be observed by dispersing anisotropic colloidal particles in an isotropic liquid [14]. A classic example is that of vanadium pentoxide, V_2O_5, which had already been shown about a century ago by Freundlich [15] to be anisotropic. Nevertheless, also many other minerals and clays, lead to inorganic liquid crystals, as has been reviewed by Sonin [16].

Thermotropic dispersions and lyotropic liquid crystalline behaviour have also been reported for carbon based materials, for example involving single-walled and multi-walled nanotubes [17–22] for electrically and magnetically addressable molecular switches. Also lyotropic graphene oxide [23–28] has been explored in a variety of host liquids for possible electro-optic applications based on the Kerr effect. Further reports discuss inorganic nanorods [29–33], ferroelectric particles [34–36] and magnetic nanorods [37,38] and platelets for ferromagnetic nematics. Also the incorporation of gold nanoparticles [39–41] into liquid crystals or indeed mesogenic molecules has become popular, especially for applications in plasmonics. Furthermore, carbon materials such as fullerenes [42,43] are also incorporated into liquid crystal forming molecules. The general reasons for dispersing colloids in liquid crystals by a variety of different methods and procedures [44,45], are to tune the liquid crystal properties, to add functionality, or to exploit the self-organization of the liquid crystal and use the order of the host as a template to transfer order onto dispersed nanomaterials.

The mechanical properties, dominated by extremely small elastic constants when compared to solid state materials, are another of the characteristics of all liquid crystals. The fact that elastic constants are very small, implies that topological defects in liquid crystals extend over large, macroscopic distances, so that they can easily be observed in polarizing microscopy. This in turn leads to textures with topological defects of strength s = ±1/2 (two-fold brushes) and s = ±1 (four-fold brushes), where defects of opposite sign and equal strength attract each other and annihilate [46–48]. Defect annihilation in liquid crystals is a means to study fundamental dynamical theories, collectively known as the Kibble-Zurek mechanism [49,50] in an elegant way. From a more practical point of view, such defect structures can be stabilized by confinement in one-, or two-dimensional arrays for optical elements [51] or to act as biological surface sensors [52].

In the theoretical work of Holger Stark [53] and the experiments of Igor Musevic et al. [54] it was shown that defects are also induced when micro-spheres are placed in a well oriented nematic liquid crystal. Different types of defects can be observed for different anchoring conditions on the micro-particles, called hedgehog and Saturn ring defects. Further, the attractive bipolar or quadrupolar force between defects can also lead to the phenomenon of chaining, forming linear chains of colloids and zigzag-shaped chains, respectively. This can also be observed for rod-shaped colloidal particles [55,56]. Even two-dimensional arrays of nanomaterials can be formed, as was also confirmed elegantly through the computer simulation of the group around Slobodan Zumer [57]. The field of colloidal interactions studied in liquid crystals was fuelled by the initial observations in the pioneering work of Poulin et al. [58], who investigated nematic-water emulsions with water droplets acting as colloidal particles and determining the force that attracts two droplet colloids [59].

It is thus clear that the fields of nanomaterials dispersed in liquid crystals, as well as that of the formation of lyotropic liquid crystal phases by dispersing anisotropic nanomaterials in isotropic host liquids will continue to grow and attract interest from a wider community. This will include the synthesis of nanoparticle containing mesogens as much as the development of novel methods of dispersion of nanomaterials in liquid crystal hosts, both thermotropic and lyotropic. Different materials will be used, carbon based nanomaterials in zero-, one-, and two dimensions, minerals and clays, and synthetic nanorods, as well as biological nanoparticles. Different functionalities will be explored, ferroelectricity, ferromagnetism, semiconductivity, chirality, quantum dots, or plasmonic properties. And experiments will be joined by theory and computer simulations to eventually produce applications which will exploit the best of all areas, liquid crystals and nanomaterials, not only to improve display applications, but also to generate novel applications in the fields of optics, sensors, medicine and related fields.

I would like to thank all authors of this Special Issue of *Nanomaterials* for their contributions and all referees for their valuable comments and suggestions, as well as the editorial office for their constant and swift support.

Funding: This research received no external funding.

Conflicts of Interest: The author declares no conflict of interest.

References

1. Lagerwall, J.P.F.; Scalia, G. (Eds.) *Liquid Crystals with Nano and Micro-Particles*; World Scientific: Singapore, 2016.
2. Collings, P.J.; Hird, M. *Introduction to Liquid Crystals: Chemistry and Physics*; Taylor & Francis: London, UK, 1997.
3. Chandrasekhar, S. *Liquid Crystals*, 2nd ed.; Cambridge University Press: Cambridge, UK, 1992.
4. Singh, S. *Liquid Crystals: Fundamentals*; World Scientific: Singapore, 2002.
5. Jones, R.A.L. *Soft Condensed Matter*; Oxford University Press: Oxford, UK, 2002.
6. Hamley, E.W. *Introduction to Soft Matter: Revised Edition*; Wiley: Chichester, UK, 2007.

7.   Kleman, M.; Lavrentovich, O.D. *Soft Matter Physics: An Introduction*; Springer: Berlin, Germany, 2003.

8.   Hirst, L.S. *Fundamentals of Soft Matter Science*; CRC Press: Boca Raton, FL, USA, 2013.

9.   Terentjev, E.M.; Weitz, D.A. *The Oxford Handbook of Soft Condensed Matter*; Oxford University Press: Oxford, UK, 2015.

10. De Gennes, P.G.; Prost, J. *The Physics of Liquid Crystals*, 2nd ed.; Clarendon Press: Oxford, UK, 1993.

11. Dierking, I. *Textures of Liquid Crystals*; Wiley-VCH: Weinheim, Germany, 2003.

12. Petrov, G. *The Lyotropic State of Matter*; Taylor & Francis: London, UK, 1999.

13. Neto, A.M.F.; Salinas, S.R.A. *The Physics of Lyotropic Liquid Crystals*; Oxford University Press: Oxford, UK, 2005.

14. Dierking, I.; Al-Zangana, S. Lyotropic Liquid Crystal Phases from Anisotropic Nanomaterials. *Nanomaterials* **2017**, *7*, 305. [CrossRef] [PubMed]

15. Diesselhorst, H.; Freundlich, H. On the double refraction of vanadine pentoxydsol. *Phys. Z.* **1915**, *16*, 419–425.

16. Sonin, A.S. Inorganic lyotropic liquid crystals. *J. Mater. Chem.* **1998**, *8*, 2557–2574. [CrossRef]

17. Dierking, I.; Scalia, G.; Morales, P.; LeClere, D. Aligning and reorienting carbon nanotubes with nematic liquid crystals. *Adv. Mater.* **2004**, *16*, 865–869. [CrossRef]

18. Dierking, I.; Scalia, G.; Morales, P. Liquid crystal–carbon nanotube dispersions. *J. Appl. Phys.* **2005**, *97*, 044309. [CrossRef]

19. Lagerwall, J.; Scalia, G.; Haluska, M.; Dettlaf-Weglikowska, U.; Roth, S.; Giesselmann, F. Nanotube alignment using lyotropic liquid crystals. *Adv. Mater.* **2007**, *19*, 359–364. [CrossRef]

20. Kumar, S.; Bisoyi, H.K. Aligned carbon nanotubes in the supramolecular order of discotic liquid crystals. *Angew. Chem. Int. Ed.* **2007**, *46*, 1501–1503. [CrossRef] [PubMed]

21. Song, W.; Kinloch, I.A.; Windle, A.H. Nematic liquid crystallinity of multiwall carbon nanotubes. *Science* **2003**, *302*, 1363. [CrossRef] [PubMed]

22. Badaire, S.; Zakri, C.; Maugey, M.; Derre, A.; Barisci, J.N.; Wallace, G.; Poulin, O. Liquid crystals of DNA-stabilized carbon nanotubes. *Adv. Mater.* **2005**, *13*, 1673–1676. [CrossRef]

23. Kim, J.E.; Han, T.H.; Lee, S.H.; Kim, J.Y.; Ahn, C.W.; Yun, J.M.; Kim, S.O. Graphene oxide liquid crystals. *Angew. Chem. Int. Ed.* **2011**, *50*, 3043–3047. [CrossRef] [PubMed]

24. Xu, Z.; Gao, C. Aqueous liquid crystals of graphene oxide. *ACS Nano* **2011**, *5*, 2908–2915. [CrossRef] [PubMed]

25. Shen, T.Z.; Hong, S.H.; Song, J.K. Electro-optical switching of graphene oxide liquid crystals with an extremely large Kerr coefficient. *Nat. Mater.* **2014**, *13*, 394. [CrossRef] [PubMed]

26. Al-Zangana, S.; Iliut, M.; Turner, M.; Vijayaraghavan, A.; Dierking, I. Properties of a thermotropic nematic liquid crystal doped with graphene oxide. *Adv. Opt. Mater.* **2016**, *4*, 1541–1548. [CrossRef]

27. Al-Zangana, S.; Iliut, M.; Turner, M.; Vijayaraghavan, A.; Dierking, I. Confinement effects on lyotropic nematic liquid crystal phases of graphene oxide dispersions. *2D Mater.* **2017**, *4*, 041004. [CrossRef]

28. Narayan, R.; Kim, J.E.; Kim, J.Y.; Lee, K.E.; Kim, S.O. Liquid Crystals: Graphene Oxide Liquid Crystals: Discovery, Evolution and Applications (Adv. Mater. 16/2016). *Adv. Mater.* **2016**, *28*, 3044. [CrossRef] [PubMed]

29. Saliba, S.; Mingotaud, C.; Kahn, M.L.; Marty, J.-D. Liquid crystalline thermotropic and lyotropic nanohybrids. *Nanoscale* **2013**, *5*, 6641–6661. [CrossRef] [PubMed]

30. Zhang, S.; Majewski, P.W.; Keskar, G.; Pfefferle, L.D.; Osuji, C.O. Lyotropic self-assembly of high-aspect-ratio semiconductor nanowires of single-crystal ZnO. *Langmuir* **2011**, *27*, 11616–11621. [CrossRef] [PubMed]

31. Ren, Z.; Chen, C.; Hu, R.; Mai, K.; Qian, G.; Wang, Z. Two-Step Self-Assembly and Lyotropic Liquid Crystal Behavior of TiO$_2$ Nanorods. *J. Nanomater.* **2012**, *2012*, 180989. [CrossRef]

32. Li, L.-S.; Walda, J.; Manna, L.; Alivisatos, A.P. Semiconductor nanorod liquid crystals. *Nano Lett.* **2002**, *2*, 557–560. [CrossRef]

33. Thorkelsson, K.; Bai, P.; Xu, T. Self-assembly and applications of anisotropic nanomaterials: A review. *Nano Today* **2015**, *10*, 48–66. [CrossRef]

34. Li, F.H.; West, J.; Glushchenko, A.; Cheon, C.I.; Reznikov, Y. Ferroelectric nanoparticle/liquid-crystal colloids for display applications. *J. Soc. Inf. Disp.* **2006**, *14*, 523–527. [CrossRef]

35. Basu, R. Soft memory in a ferroelectric nanoparticle-doped liquid crystal. *Phys. Rev. E* **2014**, *89*, 022508. [CrossRef] [PubMed]

36. Al-Zangana, S.; Turner, M.; Dierking, I. A comparison between size dependent paraelectric and ferroelectric BaTiO$_3$ nanoparticle doped nematic and ferroelectric liquid crystals. *J. Appl. Phys.* **2017**, *121*, 085105. [CrossRef]

37. Podoliak, N.; Buchnev, O.; Bavykin, D.V.; Kulak, A.N.; Kaczmarek, M.; Sluckin, T.J. Magnetite nanorod thermotropic liquid crystal colloids: Synthesis, optics and theory. *J. Colloid Interface Sci.* **2012**, *386*, 158–166. [CrossRef] [PubMed]

38. Mertelj, A.; Lisjak, D.; Drofenik, M.; Copic, M. Ferromagnetism in suspensions of magnetic platelets in liquid crystal. *Nature* **2013**, *504*, 237. [CrossRef] [PubMed]

39. Cseh, L.; Mehl, G.H. The design and investigation of room temperature thermotropic nematic gold nanoparticles. *J. Am. Chem. Soc.* **2006**, *128*, 13376–13377. [CrossRef] [PubMed]

40. Liu, Q.K.; Cui, Y.X.; Gardner, D.; Li, X.; He, S.L.; Smalyukh, I.I. Self-Alignment of Plasmonic Gold Nanorods in Reconfigurable Anisotropic Fluids for Tunable Bulk Metamaterial Applications. *Nano Lett.* **2010**, *10*, 1347–1353. [CrossRef] [PubMed]

41. Dintinger, J.; Tang, B.J.; Zeng, X.B.; Liu, F.; Kienzler, T.; Mehl, G.H.; Ungar, G.; Rockstuhl, C.; Scharf, T. A Self-Organized Anisotropic Liquid-Crystal Plasmonic Metamaterial. *Adv. Mater.* **2013**, *25*, 1999–2004. [CrossRef] [PubMed]

42. Sawamura, M.; Kawai, K.; Matsuo, Y.; Kanie, K.; Kato, T.; Nakamura, E. Stacking of conical molecules with a fullerene apex into polar columns in crystals and liquid crystals. *Nature* **2002**, *419*, 702. [CrossRef] [PubMed]

43. Lehmann, M.; Huegel, M. A Perfect Match: Fullerene Guests in Star-Shaped Oligophenylenevinylene Mesogens. *Angew. Chem. Int. Ed.* **2015**, *54*, 4110–4114. [CrossRef] [PubMed]

44. Hegmann, T.; Qi, H.; Marx, V.M. Nanoparticles in liquid crystals: Synthesis, self-assembly, defect formation and potential applications. *J. Inorg. Organomet. Polym. Mater.* **2007**, *17*, 483–508. [CrossRef]

45. Stamatoiu, O.; Mirzaei, J.; Feng, X.; Hegmann, T. Nanoparticles in liquid crystals and liquid crystalline nanoparticles. *Top. Curr. Chem.* **2012**, *318*, 331–393. [PubMed]

46. Chuang, I.; Durrer, R.; Turok, N.; Yurke, B. Cosmology in the laboratory: Defect dynamics in liquid crystals. *Science* **1991**, *251*, 1336–1342. [CrossRef] [PubMed]

47. Dierking, I.; Marshall, O.; Wright, J.; Bulleid, N. Annihilation dynamics of umbilical defects in nematic liquid crystals under applied electric fields. *Phys. Rev. E* **2005**, *71*, 061705. [CrossRef] [PubMed]

48. Dierking, I.; Ravnik, M.; Lark, E.; Healey, J.; Alexander, G.P.; Yeomans, J.M. Anisotropy in the annihilation dynamics of umbilic defects in nematic liquid crystals. *Phys. Rev. E* **2012**, *85*, 021703. [CrossRef] [PubMed]

49. Kibble, T.W.B. Topology of cosmic domains and strings. *J. Phys. A* **1976**, *9*, 1387. [CrossRef]

50. Zurek, W.H. Cosmological experiments in superfluid helium. *Nature* **1985**, *317*, 505. [CrossRef]

51. Migara, L.K.; Lee, C.-M.; Kwak, K.; Lee, H.; Song, J.-K. Tunable optical vortex arrays using spontaneous periodic pattern formation in nematic liquid crystal cells. *Curr. Appl. Phys.* **2018**, *18*, 819–823. [CrossRef]

52. Brake, J.M.; Daschner, M.K.; Luk, Y.Y.; Abbott, N.L. Biomolecular interactions at phospholipid-decorated surfaces of liquid crystals. *Science* **2003**, *302*, 2094–2097. [CrossRef] [PubMed]

53. Stark, H. Physics of colloidal dispersions in nematic liquid crystals. *Phys. Rep.* **2001**, *351*, 387–474. [CrossRef]

54. Musevic, I. *Liquid Crystal Colloids*; Springer: Cham, Switzerland, 2017.

55. Tkalec, U.; Skarabot, M.; Musevic, I. Interactions of micro-rods in a thin layer of a nematic liquid crystal. *Soft Matter* **2008**, *4*, 2402–2409. [CrossRef]

56. Dierking, I.; Heberle, M.; Osipov, M.A.; Giesselmann, F. Ordering of ferromagnetic nanoparticles in nematic liquid crystals. *Soft Matter* **2017**, *13*, 4636–4643. [CrossRef] [PubMed]

57. Ravnik, M.; Skarabot, M.; Zumer, S.; Tkalec, U.; Poberaj, I.; Babic, D.; Osterman, N.; Musevic, I. Entangled nematic colloidal dimers and wires. *Phys. Rev. Lett.* **2007**, *99*, 247801. [CrossRef] [PubMed]

58. Poulin, P.; Stark, H.; Lubensky, T.C.; Weitz, D.A. Novel colloidal interactions in anisotropic fluids. *Science* **1997**, *275*, 1770–1773. [CrossRef] [PubMed]

59. Poulin, P.; Cabuil, V.; Weitz, D.A. Direct measurement of colloidal forces in an anisotropic solvent. *Phys. Rev. Lett.* **1997**, *79*, 4862. [CrossRef]

nanomaterials

MDPI

Review

Ferroelectric Nanoparticles in Liquid Crystals: Recent Progress and Current Challenges

Yuriy Garbovskiy * and Anatoliy Glushchenko

UCCS Biofrontiers Center and Department of Physics, University of Colorado Colorado Springs,
Colorado Springs, CO 80918, USA; aglushch@uccs.edu
* Correspondence: ygarbovs@uccs.edu; Tel.: +1-719-255-3123

Received: 7 October 2017; Accepted: 24 October 2017; Published: 1 November 2017

Abstract: The dispersion of ferroelectric nanomaterials in liquid crystals has recently emerged as a promising way for the design of advanced and tunable electro-optical materials. The goal of this paper is a broad overview of the current technology, basic physical properties, and applications of ferroelectric nanoparticle/liquid crystal colloids. By compiling a great variety of experimental data and discussing it in the framework of existing theoretical models, both scientific and technological challenges of this rapidly developing field of liquid crystal nanoscience are identified. They can be broadly categorized into the following groups: (i) the control of the size, shape, and the ferroelectricity of nanoparticles; (ii) the production of a stable and aggregate-free dispersion of relatively small (~10 nm) ferroelectric nanoparticles in liquid crystals; (iii) the selection of liquid crystal materials the most suitable for the dispersion of nanoparticles; (iv) the choice of appropriate experimental procedures and control measurements to characterize liquid crystals doped with ferroelectric nanoparticles; and (v) the development and/or modification of theoretical and computational models to account for the complexity of the system under study. Possible ways to overcome the identified challenges along with future research directions are also discussed.

Keywords: liquid crystals; nanomaterials; ferroelectric nanoparticles; spontaneous polarization; aggregation; nanocolloids; electro-optics; ions

---

1. Liquid Crystals and Nanoparticles: Introduction

Nanoparticles in liquid crystals remain a hot topic of modern soft condensed matter research. This statement becomes obvious considering hundreds of published papers reviewed in multiple publications [1–5]. The rise of nanotechnology in late 1990s revitalized the idea, expressed by F. Brochard and P. G. de Gennes back in 1970, to change the properties of liquid crystals by mixing them with sub-micrometer magnetic particles [6]. Since that time various types of nanomaterials mixed with liquid crystals were studied including magnetic [7,8], ferroelectric [8,9], dielectric [8], semiconductor [8,10,11], metal [8,12], polymer [8], and carbon-based (nanotubes, fullerenes, etc.) [8,13,14] nano-dopants (for more detail please also refer to a recently published collective monograph [1]).

The dispersion of nanoparticles in liquid crystals proved to be a very fertile concept leading to the variety of interesting effects and new multifunctional materials [1–14]. Given a tremendous amount of existing literature on nanomaterials in liquid crystals, we narrowed the scope of this paper by considering liquid crystals doped with ferroelectric nanoparticles, a research topic pioneered by Y. Reznikov to whom we dedicate this topical review.

2. Liquid Crystals Doped with Ferroelectric Nanoparticles: A Brief Historical Overview

The very first paper reporting systematic studies of nematic liquid crystals doped with ferroelectric nanoparticles was published back in 2003 [15]. The major idea of the paper [15] was to increase the sensitivity of liquid crystals to the electric field and their electro-optical performance through mixing them with ferroelectric nanomaterials. To reduce the aggregation of ferroelectric nanoparticles, (i) they were functionalized with oleic acid; and (ii) their volume concentration was relatively low (<1%).

Main features of the diluted suspension of ferroelectric nanoparticles in nematic liquid crystals reported in paper [15] include (1) nearly 2-fold enhanced dielectric anisotropy; (2) nearly 2-fold lowering of the threshold voltage; (3) linear electro-optical response, or, in other words, the sensitivity of the suspension to the sign of the applied electric field, a property intrinsic to ferroelectric liquid crystals rather than to nematics. Very strong electric field generated by a ferroelectric nanoparticle along with alignment of these nanoparticles in liquid crystals were considered a major physical reason leading to the aforementioned features (1)–(3).

These findings, intriguing and very promising for applications, initiated very active research into the properties of liquid crystals doped with ferroelectric nanoparticles. Indeed, the total number of the published papers exhibits nearly linear increase during the last decade as shown in Figure 1. This figure also indicates high interest of the scientific community to this research topic.

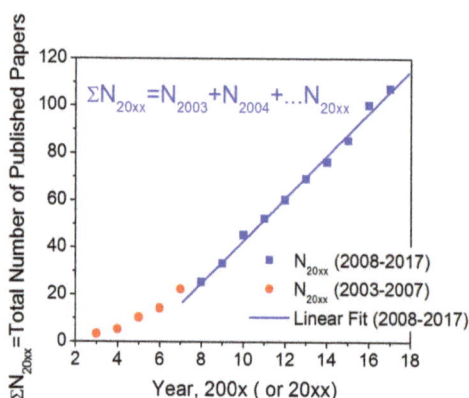

Figure 1. Total number of published papers reporting the properties of liquid crystals doped with ferroelectric nanomaterials versus time.

A distribution of published journal papers along with major research highlights over the 2003–2017 periods are schematically shown in Figure 2 (published papers: 2003—[15–17]; 2004—[18,19]; 2005—[20–24]; 2006—[25–28]; 2007—[29–36]; 2008—[37–39]; 2009—[40–47]; 2010—[48–59]; 2011—[60–66]; 2012—[67–74]; 2013—[75–83]; 2014—[84–90]; 2015—[91–99]; 2016—[100–114]; 2017—[115–123]).

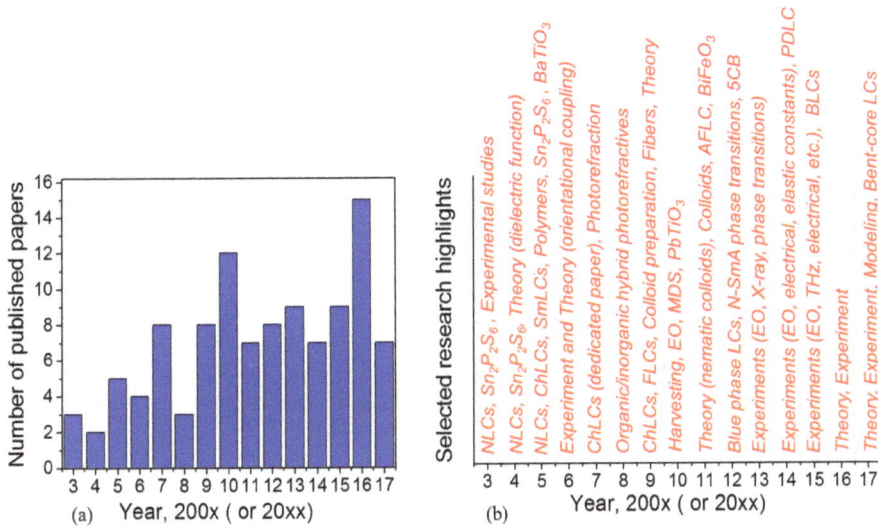

Figure 2. (**a**) Number of papers published during the 2003–2017 period; and (**b**) major research highlights. Nematic liquid crystals (NLCs), cholesteric liquid crystals (ChLCs), smectic liquid crystals (SmLCs), ferroelectric liquid crystals (FLCs), antiferroelectric liquid crystals (AFLC), polymer dispersed liquid crystals (PDLC), blue phase liquid crystals (BLCs), liquid crystals (LC), electro-optics (EO), molecular dynamics simulation (MDS).

2.1. Early Developments (2003–2006)

During the first several years (2003–2006) practically all published papers [15–28] came from the same research team (Ukraine-US-UK). The materials of choice were ferroelectric nanoparticles (SPS = $Sn_2P_2S_6$ and BTO = $BaTiO_3$) dispersed in nematics [15–28], smectics [20,26,28], and cholesterics [20,26,28]. A simplified theory of ferroelectric nanoparticle/liquid crystal colloids was developed to explain experimental results [19,25,27].

2.2. Research Expansion (2007–2011)

Starting from 2007 more and more research groups became involved in studying the properties of liquid crystals doped with ferroelectric nanoparticles (Figure 2, [29–66]). At the same time, the scope of research interests significantly expanded. In addition to dielectric properties, electro-optics, and phase transitions thoroughly investigated during the 2003–2006 time period [15–28], photorefractive phenomena in organic–inorganic hybrids [32,33,35,37,38,46,50], polarization fluctuations observed in such systems [36], the effects of the nanoparticle size [56], and applications of liquid crystal/ferroelectric nanoparticle colloids for the design of electro-optical devices [31,39], tunable fibers [45] and alignment layers [44] were reported (see also Figure 2). The very first papers exclusively focused on ferroelectric smectic liquid crystals [41,55,66] and cholesteric liquid crystals [30,42] doped with ferroelectric nanoparticles also appeared.

Further progress in a theory of liquid crystals doped with ferroelectric nanoparticles was also made by considering how the orientational order of liquid crystals was affected by the orientational order of nanoparticles [47], by introducing Maier-Saupe-type theory of ferroelectric nanoparticles in nematic liquid crystals [65], and by analyzing Freedericksz transition in ferroelectric liquid-crystal nanosuspensions [64]. In addition, molecular dynamics simulations of spherical ferroelectric nanoparticles immersed in nematic liquid crystals were also performed [51].

Around the same time it became obvious that the properties of liquid crystals doped with ferroelectric nanoparticles depend strongly on the way ferroelectric nanoparticles are prepared [40,43,63,124]. As a result, the preparation of ferroelectric [40] and paraelectric [43] nanoparticles for their use in liquid crystal colloids and their applications was also thoroughly discussed. However, liquid crystals doped with ferroelectric nanoparticles turned out to be very delicate systems. As was shown in paper [48], even single component liquid crystals (5CB) doped with ferroelectric nanoparticles (SPS = $Sn_2P_2S_6$) could exhibit different behavior such as increase and decrease in the threshold voltage and the "nematic-isotropic" phase transition temperature. Many factors could cause this very complex behavior [8,48,125]. For example, the electric field of nanoparticles could be screened by the charges in liquid crystals; nanoparticles could lose their ferroelectricity during their preparation; effective "dilution" of liquid crystals by nanoparticles, etc. [126]. These multiple factors could mask effects expected for liquid crystals doped with ferroelectric nanoparticles thus leading to the reported effects [8,125].

The ferroelectricity of nanoparticles was considered the major reason leading to the increased order parameter and the modification of the physical properties of the liquid crystal host [8,25,47,51,64,65]. That is why the need for the development of experimental methods allowing the production of truly ferroelectric nanoparticles suitable for their dispersion in liquid crystals became very urgent. An elegant method to harvest ferroelectric nanoparticles was reported in paper [49]. This technique was applied to study holographic beam coupling in inorganic-organic photorefractive hybrids using liquid crystals doped with harvested nanoparticles [50]. Moreover, the use of harvested ferroelectric nanoparticles revealed asymmetric Freedericksz transitions in symmetric liquid crystal cells doped with such harvested nanoparticles [57].

Since 2007, the choice of liquid crystals and ferroelectric nanoparticles was also gradually expanding by introducing new materials to study. For example, lead titanate (PTO) nanoparticles embedded in a liquid crystalline elastomer matrix and multiferroic $BiFeO_3$ nanoparticles dispersed in partially fluorinated orthoconic antiferroelectric liquid crystal were studied in papers [54,66], respectively.

2.3. Research Expansion, Globalization, and Validation (2012–2017)

Research into the properties of liquid crystals doped with ferroelectric nanoparticles, initially undertaken by European and American scientists mostly, received a considerable boost due to the contributions coming from China, India, Iran, Taiwan, South Korea, and Japan. Several papers published in 2010–2011 [53,55,61,66] were followed up by an even greater number of publications [67,70,78,79,81,88,97,100,102,104,107,110,111,116,117,119]. For example, low voltage and hysteresis free blue phase liquid crystals doped with ferroelectric nanoparticles were reported [67,97]. Interesting effects of ferroelectric nanoparticles on the luminescence and electro-optics of ferroelectric liquid crystals were observed [78,81,85,89,90,95,96,105–109], and theory of (i) nanoparticles in ferroelectric liquid crystals [79]; (ii) the effect of ferroelectric nanoparticles on the isotropic-smectic-A phase transition [104] and the Freedericksz transition in smectic-A liquid crystals [119]; and (iii) the dielectric permittivity in the isotropic phase of the isotropic-smectic-A phase transition were also developed [117].

During the 2012–2017 time period the variety of materials used in experimental studies continued to grow. Polymer stabilized blue phase liquid crystals [97], polymer dispersed liquid crystals [88,110,111], and bent-core liquid crystals [122] doped with barium titanate (BTO) nanoparticles were studied. Moreover, lithium niobate (LNO) and multi-ferroic nanoparticles were introduced as ferroelectric dopants [85,99].

The effects of ferroelectric nanoparticles on the properties of a single component liquid crystal such as 5CB still remain among major research interest [72,107,108]. Electro-optics (including the Freedericksz transitions in nematic and smectic-A liquid crystals) [98,119]; dielectric [85,102,105,118,123], electrical [85,86,92,93,100,102,103,107,116,121,122], and viscoelastic [90,96,122] properties; phase transitions and pre-transitional effects [80,94–96,107,112,113,118,122,123] in liquid crystals doped with ferroelectric nanoparticles are also receiving due attention during this time period. Another new direction gaining interest of the scientific community includes studies of hybrid liquid crystal-ferroelectric nanoparticles composite materials in the terahertz and microwave regions [91,120].

A very important aspect of current research is an increasing use of the harvested nanoparticles in experimental studies with the goal to distinguish between direct effects of the nanoparticle's ferroelectricity on the properties of liquid crystals and concurrent effects such as the dilution effect, screening effect, etc. [77,85,94,105,112,118,123].

3. Technology and Basic Properties of Liquid Crystals Doped with Ferroelectric Nanoparticles

3.1. Current Technology

Liquid crystals doped with ferroelectric nanoparticles are typically produced by mixing a small amount of nanomaterials with the liquid crystal host. Prior to making the liquid crystal nano-dispersion, nanoparticles can be either functionalized with the surfactant (in this case they are typically dispersed in an isotropic and non-ionic liquid carrier such as heptane or toluene) or they can be in the form of nano-powders (without any capping agent). The surfactant is needed to reduce the aggregation of nanoparticles. So far, oleic acid is the most widely used surfactant. However, this surfactant is not an ideal option since the affinity of its molecules to the surface of nanoparticles is rather moderate [125].

Once nanoparticles are dispersed in liquid crystals, the obtained dispersion is subject to sonication and/or prolonged shaking to mediate the tendency of nanoparticles to aggregate, and to allow the non-ionic liquid containing nanoparticles for its complete evaporation from the liquid crystal mixture.

To date, the major method to produce ferroelectric nanoparticles for their applications in liquid crystal nano-colloids is a mechanical grinding of ferroelectric materials [8,40,125]. The ferroelectric material is milled together with the surfactant in a dielectric fluid carrier. The size of the obtained nanoparticles and their size distribution is governed by the milling time which strongly depends on the choice of materials and the design of the milling machine [8,40,125]. An example of the commercially available ball mill along with the "size vs. time" dependence is shown in Figure 3.

The "harvesting procedure" proposed in [49] is a very important step in the preparation of true ferroelectric nanoparticles by means of ball milling. The idea of this technique is to separate ferroelectric and non-ferroelectric nanoparticles by subjecting the dispersion of nanoparticles to a strongly non-homogeneous electric field. In this case nanoparticle with a non-zero permanent dipole moment will move whereas nanoparticles with net zero dipole moment will not "feel" the gradient of the electric field [49]. The schematic of the harvesting setup is shown in the inset, Figure 3d. A high DC (direct current) voltage (typically 10–20 kV) is applied across the inner wire electrode and the outer cylindrical electrode which is typically grounded. During the harvesting procedure, the harvested nanoparticles with permanent dipoles are accumulated on the inner wire electrode whereas non-ferroelectric nanoparticles remain in the fluid harvesting medium. Recent experiments with harvested ferroelectric nanoparticles provide new insights into the effects of their intrinsic ferroelectricity on the properties of liquid crystals [77,85,96,101,123]. It should be noted that ideal harvesting assumes non-charged ferroelectric nanoparticles. However, some ferroelectric nanomaterials such as SPS can be charged thus significantly complicating the harvesting process and interpretation of the obtained experimental results [92,125].

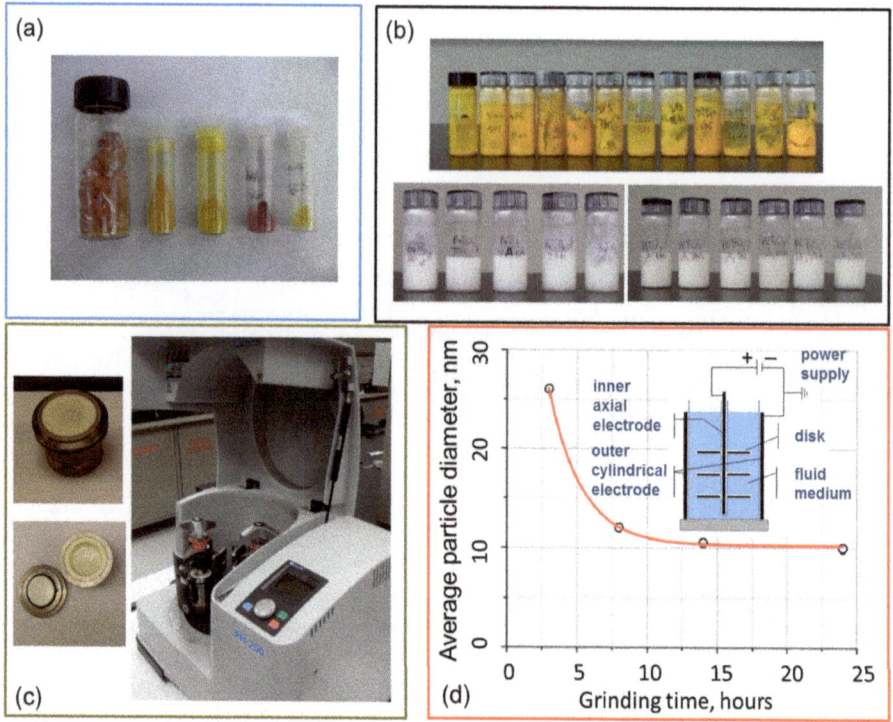

Figure 3. (**a**) Macro-crystals prior to milling; (**b**) The obtained dispersion of milled ferroelectric nanoparticles in a fluid carrier (heptane); (**c**) Commercially available high-energy ball mill; (**d**) an average size of the milled nanoparticles vs. grinding time [8]. A typical harvesting setup is shown in the inset (redrawn after [125]).

3.2. Basic Properties of Liquid Crystals Doped with Ferroelectric Nanoparticles

Consider ferroelectric nanoparticle immersed in a liquid crystal host. An electric field, $\vec{E}$, in the vicinity of ferroelectric nanoparticle approximated as a dielectric sphere with a spontaneous polarization, $\vec{P_S}$, can be estimated according to Equation (1) [125]:

$$\vec{E} = \frac{R_{NP}^3}{3\varepsilon_0\varepsilon} \left(\frac{3\left(\vec{P_S} \cdot \vec{r}\right)\vec{r}}{r^5} - \frac{\vec{P_S}}{r^3} \right) \tag{1}$$

where R_{NP} is a radius of nanoparticle, ε_0 is an electric constant, ε is a dielectric permittivity of the surrounding medium, and $\vec{r}$ is a radius vector. The value of this field is very high, on the order of 10^9 V/m. This field can be comparable to or even stronger than the field due to intermolecular interactions in liquid crystals. Therefore, ferroelectric nanoparticles can directly affect the orientation of liquid crystal molecules in the vicinity of nanoparticles thus resulting in the field coupling between the liquid crystal director and the dipole moment of nanoparticles. In addition to the coupling with the liquid crystal director, this strong field can also change intermolecular interactions in the vicinity of nanoparticles [25,125]. These two factors can result in the increased ordering of liquid crystal molecules characterized by the order parameter *S*.

An increase in the order parameter changes basic properties of liquid crystals determined by S. The most important ones include the birefringence ($\Delta n \propto S$), dielectric anisotropy ($\Delta n \propto S$), and elastic constants ($K \propto S$). Consequently, these changes should strongly affect the threshold voltage (the Freedericksz transition) of the liquid crystal nano-colloid [125]. In addition, an increased ordering of liquid crystals doped with ferroelectric nanoparticles can lead to the increase in the clearing temperature, T_c. More details on theory supporting these conclusions can be found in papers [25,47,65,84,87]. Here we just provide a brief summary of major theoretical results.

The effect of ferroelectric nanoparticles on the dielectric anisotropy and Freedericksz transition in nematic liquid crystals was analyzed in papers [19,27,64]. Assuming strong nanoparticle–liquid crystals coupling, the obtained mathematical expression for the threshold voltage was found to be similar to a standard expression for a pure nematic:

$$V_{Fr} = \pi \sqrt{\frac{K}{\varepsilon_0 \Delta \varepsilon_{eff}}} \tag{2}$$

where K is an elastic constant, $\Delta \varepsilon_{eff}$ is an anisotropy of the effective dielectric constant of the dispersion. Due to the spontaneous polarization of ferroelectric nanoparticles, the effective permittivity of liquid crystals/nanoparticles dispersion along the director is larger than the permittivity of pure liquid crystals. According to [64], a small volume fraction of ferroelectric nanoparticles (on the order of a few percent) is enough to reduce the threshold voltage by a factor of two.

An increase in the clearing temperature (ΔT_c) of nematic liquid crystals doped with ferroelectric nanoparticles was analyzed in several papers [25,47,65]. Assuming perfect alignment of all ferroelectric nanoparticles along the liquid crystal director and applying the Maier-Saupe theory, the following expression was obtained [25]:

$$\Delta T_c = \frac{z f_v (\Delta \beta)^2}{163.44 \pi \varepsilon_0 l_{m-m}^3 k_B} P_S^2 \tag{3}$$

where z is the nearest number of neighboring molecules separated by a distance l_{m-m}, f_v is the volume fraction of nanoparticles, $\Delta \beta$ is the anisotropy of the molecular polarizability, $k_B = 1.38 \times 10^{-32} J/K$.

The consideration of the orientational distribution of nanoparticles in liquid crystals through the Landau theory along with the concept of coupled orientational order parameters for the liquid crystals and nanoparticles resulted in Formula (4) [47]:

$$\Delta T_c = \frac{f_v \Delta \varepsilon}{135 \rho_{LC} k_B \varepsilon_0 \varepsilon^2} P_S^2 \tag{4}$$

The generalization of the Landau theory [47] presented in paper [65] yielded another expression for the increase in the clearing temperature of nanocolloids:

$$\Delta T_c = \frac{\pi f_v R_{NP}^3}{3 T_c \rho_{LC}} \left(\frac{2 \Delta \varepsilon P_S^2}{675 k_B \varepsilon_0 \varepsilon^2} \right)^2 \tag{5}$$

In all cases (3)–(5) an increase in the clearing temperature is proportional to the concentration of nanoparticles. Both expressions (3) and (4) exhibit a quadratic dependence, $\Delta T_c \propto P_S^2$, whereas according to Formula (5) $\Delta T_c \propto P_S^4$. It should be noted that this scenario ($\Delta T_c \propto P_S^4$) was predicted assuming weak-interaction regime whereas expression (4) was obtained for the strong interaction regime (nanoparticles are small enough and the liquid crystal alignment is not distorted) [65].

4. Scientific and Technological Challenges

Early experimental and theoretical works, revealing an increased clearing temperature, enhanced order parameter, increased dielectric anisotropy and birefringence of liquid crystals doped with ferroelectric nanoparticles, stimulated very active research around the globe (Figures 1 and 2). The reported findings were very promising from both academic and practical aspects. In fact, they showed a path toward a non-synthetic design of new liquid crystal materials by means of ferroelectric nanoparticles [8,20,21,26,31,125]. However, an expansion and globalization of the research into the properties of liquid crystals doped with ferroelectric nanoparticles identified many problems resulting in a poor reproducibility of the reported results. As an example, we compiled experimental data reported by independent research groups for the same single component liquid crystals (5CB) doped with ferroelectric nanoparticles (Table 1).

According to Table 1, even in the case of the same liquid crystals (5CB) the reported results [26,48,52,75,96,98,107,108] vary substantially. For example, there are papers reporting both increase [26,48,52] and decrease in the clearing point [48,96,107,108] of 5CB liquid crystals through doping them with ferroelectric nanoparticles. This variability of experimental data depends strongly on the material parameters of ferroelectric nanoparticles used in experiments. In some cases [75,107,108], the ferroelectricity of nanomaterials was not confirmed or checked. As a result, this type of experimental data is incomplete and prevents us from making any conclusions regarding possible effects of the nanoparticle ferroelectricity on the properties of liquid crystals. The size of nanoparticles also matters: the use of very large nanoparticles disturbs the director field [108]. Moreover, the prepared colloids are prone to aggregation and are very unstable. Aging phenomena and prehistory of the prepared ferroelectric nanomaterials are also important factors [48].

An analysis of Table 1 is very useful and instructive in identifying common challenges faced by scientists. These challenges can be broadly categorized into the following groups: (i) the control of the size, shape, and the ferroelectricity of nanoparticles; (ii) the production of a stable and aggregate-free dispersion of relatively small (~10 nm) ferroelectric nanoparticles in liquid crystals; (iii) the choice of ferroelectric nanomaterials and the selection of liquid crystals the most suitable for the dispersion of nanoparticles; (iv) the choice of appropriate experimental procedures and control measurements to characterize liquid crystals doped with ferroelectric nanoparticles; and (v) the development and/or modification of theoretical and computational models to account for the complexity of the system under study.

Let us discuss each of these groups in more detail.

Table 1. Single component nematic liquid crystals (5CB) doped with ferroelectric nanoparticles

Studied Samples	Observed Effects	Reference
Quasi-spherical (20 ± 10 nm) ferroelectric nanoparticles ($Sn_2P_2S_6$, ~0.3 vol. %) were dispersed in 5CB. Nanoparticles were prepared by means of mechanical wet grinding. To provide the stability of nano-colloids, oleic acid was used as surfactant.	Several samples were prepared. The obtained results were dependent on the pre-history of the sample indicating possible aging of ferroelectric dispersions. As a result, both increase and decrease of the order parameter S (on the order of 5–10%) and of the clearing point T_c (on the order of 1–10 deg.) was demonstrated.	[48]
Ferroelectric nanoparticles ($BaTiO_3$; 1–4 vol. %; ~150 nm) were dispersed in 5CB. Nanoparticles were prepared by means of mechanical wet grinding. To provide the stability of nano-colloids, oleic acid was used as a surfactant.	An increase in the clearing point, T_{NI}, from 35.2 °C to 36.6 °C. The threshold voltage (the Freedericksz transition, V_{Fr}) was reduced from 0.79 V to 0.54 V. The order parameter was increased from 0.55 to 0.60. The turn-on time is decreased from 450 ms to 300 ms whereas the turn-off time is increased from 5.26 s to 7.75 s.	[52]
Ferroelectric nanoparticles ($BaTiO_3$; ~1 wt. %; 30–50 nm) and ($Sn_2P_2S_6$, ~200 nm) were dispersed in 5CB. Nanoparticles were prepared by means of mechanical wet grinding. To provide the stability of nano-colloids, oleic acid was used as a surfactant.	Signigicant (~2-fold) increase of the dielectric constants and 10–20% increase in the birefringence	[26]
Nanoparticles ($BaTiO_3$; ~0.5 wt. %; ~4–40 nm) were dispersed in 5CB. Nanoparticles were prepared by means of mechanical wet grinding. The ferroelectricity of nanoparticles was not confirmed by experiments. To provide the stability of nano-colloids several surfactants including oleic acid were used.	An apparent shift of the Freedericksz transition towards a slightly higher value (according to electro-optical measurements) was not confirmed by capacitance measurements. The use of surfactants made of "nematogenic" molecules results in much more stable suspensions as compared to the use of oleic acid.	[75]
Ferroelectric nanoparticles ($BaTiO_3$; 0.33–0.50 vol. %; ~50 nm) were dispersed in 5CB.	The threshold voltage (the Freedericksz transition, V_{Fr}) was reduced from 0.64 V to 0.56 V (0.33 vol. %) and 0.51 V (0.50 vol. %).	[98]
$BaTiO_3$ nanoparticles (~100 nm; 0.05–5 wt. %) were dispersed in 5CB. No data on the ferroelectricity of the dispersed nanoparticles.	A decrease in the clearing point, T_c (by about ~2 °C) was observed. The nematic temperature range is shortened with an increase in the concentration of nanoparticles. A decrease in the the dielectric anisotropy (from 13.1 to 11.2). The reduction of the threshold voltage from 1.02 V to 0.94 V. The splay elastic constant (K_{11}) is decreased from 16.50 pN to 11.13 pN (0.05 wt. %), 7.91 pN (0.5 wt. %), and to 8.88 pN (5.0 wt. %). Both increase (~100 times, at 0.05 wt. %) and decrease (~10 times, at 5.0 wt. %) in the electrical conductivity measured along the director was observed.	[107]
Ferroelectric nanoparticles ($BaTiO_3$; ~0.2–0.4 wt. %; ~12 nm) were dispersed in 5CB. Nanoparticles were prepared by means of mechanical wet grinding and harvested. To provide the stability of nano-colloids (over a few months), oleic acid was used as a surfactant.	A decrease in the clearing point, T_c (by about 2.5 °C) was observed. The enthalpy of this transition ($\Delta H_{NI} \approx 3 \, J/g$) remains almost unchanged. The nematic temperature range is shortened with an increase in the concentration of nanoparticles. Practically no change in the birefringence and the dielectric anisotropy. The splay elastic constant (K_{11}) is practically not affected by nanoparticles while the bend elastic constant(K_{33}) decreases (~20%). The decrease (~20%) in the rotational viscosity γ_1.	[96]
Relatively large $BaTiO_3$ particles (~600 nm; ~1 wt. %) were dispersed in 5CB (oleic acid was used as a surfactant).	A decrease in the clearing point, T_{NI}, from 35.2 °C to 32.4 °C, and in the rotational viscosity γ_1 from 0.081 Pa·s (pristine liquid crystals) to 0.078 Pa·s (liquid crystal nanocolloids). The reduction of the Freedericksz transition from 1.3 V to 0.3 V. The switching time: the turn-on time is decreased (~10%) whereas the turn-off time is increased (>50%).	[108]

13

4.1. Issues Related to Nanoparticles

Major issues associated with ferroelectric nanodopants include the control and evaluation of their size, shape, and the ferroelectricity.

As was already mentioned, the mechanical grinding of ferroelectric materials is a major technique to produce ferroelectric nanoparticles for their applications in liquid crystal colloids [8,40,125]. Using this method, the size of nanoparticles can be controlled by varying the milling time (Figure 3). The mechanical grinding is very delicate technique since the milling parameters should be optimized depending on the type of the mill and materials used (fluid carrier, surfactant, and ferroelectric). As a result, the produced ferroelectric nanoparticles should be carefully characterized prior to mixing them with liquid crystals.

Alternative methods to produce ferroelectric nanoparticles include physical (laser ablation, electrospinning); chemical (solid-state reaction; sol-gel technique; solvothermal method; hydrothermal method; molten salt method, and others); biological (biosynthesis) methods and their various combinations (please refer to a recent review [9] for more detail). These techniques could be very useful considering their potential to control both the size and shape of ferroelectric nanoparticles [9]. All these methods can be used in conjunction with the harvesting technique which is also very useful for the selection of truly ferroelectric nanoparticles.

To achieve good quality dispersion, relatively small ferroelectric nanoparticles are preferred. If the size of nanoparticle R_{NP} is much less than the ratio $\frac{K}{W}$ (K is the elastic constant and W is the anchoring energy) then this nanoparticle does not disturb the liquid crystal director. Instead, the dispersed ferroelectric nanoparticles modify liquid crystals' physical properties: temperature of liquid crystal phase transitions, birefringence, dielectric permittivity and magnetic permeability (if nano-multiferroics are used), conductivity, rotational viscosity, Freedericksz threshold, switching time etc. [125]. It should be noted that the size of ferroelectric nanomaterials can be reduced only to a certain critical value. Below this value ferroelectric materials typically lose their ferroelectric properties [101]. There is no universal agreement on the magnitude of this critical size since it depends on many factors including methods used to produce nanoparticles [101,127]. That is why ferroelectric properties of the newly produced nanoparticles should always be assessed by experimentalists.

Theoretical considerations described in previous section assume that nanoparticles are ferroelectric. If their ferroelectricity is not established in particular experiments, the obtained results cannot be used to verify existing theoretical predictions. Non-ferroelectric nanoparticles can affect the properties of the liquid crystal host through the dilution/anchoring effect [125,126]. For example, the change in the clearing temperature through the dilution/anchoring effect can be expressed as (6):

$$\Delta T_c = -(1+B)f_v T_c^0 \qquad (6)$$

where B is the anchoring parameter, and T_c^0 is the clearing temperature of pure (non-doped) liquid crystals [126]. Pure dilution effect ($B = 0$) can cause the reduction of the clearing temperature and the order parameter [126]. In the case of ferroelectric nanoparticles both competing effects (the effect of the nanoparticle's polarization (Equations (3)–(5)) and the dilution/anchoring effects (Equation (6)) should be considered [112,118,123].

4.2. Stable and Aggregate-Free Dispersions

The production of a stable and aggregate-free dispersion of relatively small (~10 nm) ferroelectric nanoparticles in liquid crystals still remains a major challenge experimentalists are facing nowadays.

Strong Coulomb interactions between ferroelectric nanoparticles lead to their aggregation, and the presence of this aggregation was noted in many experimental works [8,125]. Even if the produced liquid crystal nano-colloids look stable and do not show any visible sign of aggregation, small aggregates can exist [8,125]. In this case the average size of aggregates is less than the wavelength of light. The aggregation reduces the effective spontaneous polarization of the colloid, change the effective concentration of nanoparticles, and alter the stability of the colloid making its macroscopic properties time dependent. As a result, experimental data and theoretical predications are not easy to compare.

To reduce the aggregation of ferroelectric nanoparticles, they are typically capped with a surfactant and their concentration in liquid crystals is kept at a relatively low level ($f_v \ll 10^{-2}$). Up to date, the oleic acid is the most widely used surfactant. As was already mentioned, this surfactant is not an optimized choice because of its low affinity to the surface of nanoparticles. As a result, some fraction of surfactants can be dissolved in liquid crystals causing the disordering of liquid crystals and contributing to the dilution effect. This effect can be strong enough to mask the effects due to ferroelectric nanoparticles. That is why it is a good laboratory practice to study possible effects of the used surfactant on the properties of liquid crystals.

Currently the search for better surfactants is underway [125]. Very promising are mesogenic surfactants recently applied to stabilize quantum dots and produce true dispersion of semiconductor nanoparticles in liquid crystals [128]. To achieve a stable dispersion of nanoparticles in liquid crystals, a mixture of an alkyl phosphonate and a dendritic surfactant containing promesogenic units was used [128]. This enabled the formation of thermodynamically stable colloids. The minimization of the distortion of the liquid crystal ordering around the nanoparticle was considered the major reason for the achieved stability (the shelf life of the prepared colloids was longer than 1 year).

4.3. Issues Associated with the Choice of Guest-Host Materials

So far, the suitability of particular ferroelectric nanomaterials for their dispersion in liquid crystals is practically not discussed in the literature. The majority of experimental studies were done using standard ferroelectric nanomaterials (SPS = $Sn_2P_2S_6$, BTO = $BaTiO_3$, LNO = $LiNbO_3$) dispersed in nematic liquid crystals mostly (Figure 2). There are only a few publications reporting effects of multiferroic nanoparticles on the properties of liquid crystals [66,99]. Nematics are still the most widely used liquid crystal material (Figure 2). However, there is a tendency to broaden the number of liquid crystal materials by including cholesterics [30,42], different types of smectics [41,55,58,66,71,77,78,81, 82,85,105,117,119], blue phase liquid crystals [67,97], polymer dispersed liquid crystals [88,110,111], and even bent-core liquid crystals [122]. One reason for this tendency is substantial improvements in the electro-optical performance of liquid crystal materials through doping them with ferroelectric nanoparticles. Faster electro-optical switching was reported for smectic liquid crystals doped with ferroelectric nanoparticles [77,78,85,105]. In the case of cholesteric liquid crystals, a minor quantity of ferroelectric nanoparticles can cause a 45% decrease of the driving voltage along with a two-fold increase in the effective dielectric anisotropy [42]. By mixing polymer-stabilized blue phase liquid crystals with ferroelectric nanoparticles, the vertical driving electric field was dramatically (~70%) reduced [97]. An enhancement of frequency modulation response time for polymer-dispersed liquid crystal doped with ferroelectric nanoparticles was also reported [110,111].

It should be noted that in practically all reported studies thermotropic liquid crystals are typically the material of choice. The use of lyotropic liquid crystals seems to be problematic since the highly polar molecules of water will inevitably screen the electric field of ferroelectric nanoparticles (for more detail regarding the electrical behavior of ferroelectric nanoparticles in aqueous medium we refer to [129]).

Multiphase thermotropic liquid crystals exhibit different types of mesophases which are typically achieved by changing the temperature of the sample. If this is the case, the temperature range of the mesophase should be below the Curie point of the ferroelectric nano-dopant. Ferroelectric nanoparticles lose their ferroelectricity above the Curie point. If the temperature of experimental samples is higher than the Curie point, the observed effects cannot be associated with the ferroelectricity of nanoparticles.

While selecting ferroelectric nanomaterials and liquid crystals for experimental studies, their possible contamination with ions should always be considered. Mobile ions always present in liquid crystals can screen (partially or even completely) the electric field of the ferroelectric nanoparticle.

As a result, such "screened" ferroelectric nanoparticles will behave as nanodopants with significantly reduced (or even zero) effective permanent polarization. Consequently, an interpretation of the obtained experimental results and their comparison with theoretical predictions can become very problematic [125].

Strong Coulomb interactions between ions and ferroelectric nanoparticles dispersed in liquid crystals results in the well-known ion-trapping effect [2]. This effect, reported by many research groups, is very promising for the permanent purification of liquid crystals from ions. However, if ferroelectric nanoparticles are contaminated with ions prior to mixing them with liquid crystals, the observed effects can differ [130,131]. In fact, depending on the interplay between the adsorption-desorption processes in the liquid crystal nano-colloids and the level of the nanoparticle contamination, different regimes can be achieved: ion trapping regime [85,86,93], ion releasing regime [100], and no change in the concentration of ions [130]. Ferroelectric nanomaterials are very prone to uncontrolled ionic contamination. That is why it is important to assess the level of this contamination.

4.4. Experimental Procedures and Control Measurements

The choice of appropriate experimental procedures and control measurements to characterize liquid crystals doped with ferroelectric nanoparticles is not a trivial task. It can strongly affect the measured data. The electric field, originated from the spontaneous polarization of ferroelectric nanoparticles, and its interaction with surrounding mesogenic molecules is considered the major physical factor determining the properties of liquid crystal nano-colloids [125]. Therefore, an evaluation of this field through experimental assessment of the spontaneous polarization of nanoparticles is very important [63,125]. Electrical characterization of ferroelectric nanoparticles including measurements of the polarization switching current and ferroelectric hysteresis loop can provide enough information about the ferroelectricity of nanoparticles, possible screening effects, and the electric field due to the spontaneous polarization [63,124]. Complementary measurements such as Raman spectroscopy and X-ray structure analysis can identify the tetragonal structure of ferroelectric nanoparticles. However, these methods do not "see" the screening of the spontaneous polarization by mobile charges typically present in colloids.

Electrical and dielectrical measurements of liquid crystals doped with ferroelectric nanoparticles should be taken and interpreted with a high degree of caution. In the case of nematics, prior to applying the electric field, the net macroscopic polarization of the colloids is typically nearly zero (ferroelectric nematic suspensions exhibit behavior typical for paraelectrics). The applied electric field aligns dipoles of nanoparticles causing the amplification of the nanoparticle polarization. This amplification can affect the order parameter, phase transition temperatures, and the measured dielectric permittivity of the samples [112,118,123]. At the same time, techniques not utilizing electric field can yield different results for the same measured physical quantities. For example, the phase transition temperatures measured by means of dielectric spectroscopy were higher than that measured through differential scanning calorimetry [112,118]. Depending on its amplitude and frequency, the applied electric field can also alter the aggregation dynamics in liquid crystals doped with ferroelectric nanoparticles [60,63,124].

The use of control samples in experiments is very important. Measurements performed with dispersions of nanoparticles in liquid crystals should be repeated with pure (non-doped) liquid crystals under identical conditions. In the case of optical and electro-optical studies, the use of the twin cells is very beneficial (Figure 4) [125]. The twin cell is the cell artificially divided by a polymer stripe into two identical regions. These regions are characterized by the same thickness, boundary conditions and anchoring strength. One of these regions is filled with liquid crystals doped with ferroelectric nanoparticles and the other region is filled with pure liquid crystals. Optical and electro-optical measurements performed with both regions can immediately reveal an impact of ferroelectric nanoparticles on the properties of liquid crystals (Figure 4).

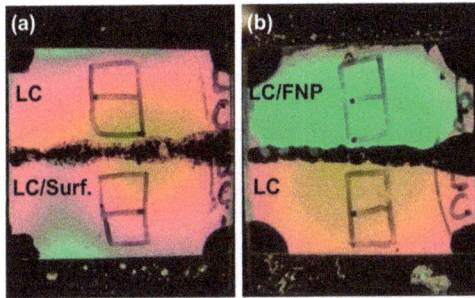

Figure 4. The twin cell placed in between two crossed polarizers: (**a**) the cell is filled with pure liquid crystals (this region is marked as "LC") and liquid crystals doped with surfactant (oleic acid) (this region is marked as "LC/Surf."); (**b**) the cell is filled with pure liquid crystals (marked as "LC") and liquid crystals doped with ferroelectric nanoparticles (marked as "LC/FNP").

4.5. New Theoretical and Computational Models

Existing theories [19,25,47,64,65,84,87,117,119] and computational models [51] considered rather simplified systems. Typically, the modeled system assumes mono-domain, neutral, and non-interacting ferroelectric nanoparticles dispersed in single component liquid crystals. At the same time, real systems are much more complex than the modeled ones. First of all, ferroelectric nanoparticles can vary in their size and the value of the spontaneous polarization. In addition, nanoparticles can aggregate and form dimers, chains, etc. In addition, nanoparticles are typically functionalized with the surfactant. Due to the finite affinity of the surfactant to the surface of nanoparticles, a fraction of this surfactant will also be dissolved in the liquid crystal host. The majority of liquid crystals are multi-component mixtures. As a result, there is a possibility for macro-nano separation of the liquid crystal components caused by ferroelectric nanoparticles. Another important aspect includes mobile charges in liquid crystals and the possibility of charged ferroelectric nanoparticles [92,125]. In addition, ferroelectric nanoparticles can be contaminated with ions prior to mixing them with liquid crystals [130,131]. All presented examples unambiguously illustrate how complex and delicate liquid crystals doped with ferroelectric nanoparticles are. More advanced theories and computational models should take into account the above-mentioned factors to ensure further progress in this very vibrant research.

5. Conclusions

Research into the properties of liquid crystals doped with ferroelectric nanoparticles has been carried out for more than a decade. Despite the broad variety of the results we can state that basic physics of *ferroelectric, non-charged and non-aggregated*, nanoparticles in *nematic* liquid crystals is reasonably understood. Ferroelectric nanoparticles do modify the properties of liquid crystals leading to the increase in the order parameter and resulting in improved electro-optical characteristics. This improved electro-optical performance (a higher birefringence, a shorter switching time, better contrast and lower threshold voltage, enhanced nonlinear-optical properties) is very promising for the design of the next generation devices which are cheaper, faster and better than currently existing products. These applications can include advanced displays, fast electro-optical switchers and shutters, memory cells, tunable filters, nonlinear-optical valves for optical processing systems, etc.

Additional studies are needed to explore a full potential of ferroelectric nano-dopants in cholesterics, various types of smectics, blue phase and bent core liquid crystals, and polymer dispersed liquid crystals. So far, among these materials, smectic liquid crystals are the mostly studied [41,55,58,66,71,77,78,81,82,85,105,117,119]. Only very limited number of papers considered cholesteric [30,42], blue phase [67,97], polymer dispersed [88,110,111] and bent core [122] liquid crystals doped with ferroelectric nanoparticles. Nevertheless, the available findings also indicate positive improvements in the properties of the aforementioned liquid crystals through doping them with ferroelectric nanoparticles.

Further progress in this vibrant research field will be achieved by introducing new experimental and theoretical concepts to solve major scientific and technological challenges discussed in the previous section. We hope our brief review will encourage students and professional researchers to "join the club" and actively participate in exploring new horizons of liquid crystals doped with ferroelectric nanoparticles, an exciting research direction launched by Prof. Y. Reznikov in early 2000s.

Acknowledgments: The authors are grateful to all their colleagues participating in a quest to unravel the mystery of liquid crystals doped with ferroelectric nanoparticles.

Author Contributions: Y.G. and A.G. conceived and designed the layout of the review, analyzed existing literature and wrote the paper.

Conflicts of Interest: The authors declare no conflict of interest.

References

1. Lagerwall, J.P.F.; Scalia, G. *Liquid Crystals with Nano and Microparticles (Series in Soft Condensed Matter: Volume 7)*; World Scientific Publishing Co.: Singapore, 2016; pp. 461–920.
2. Garbovskiy, Y.; Glushchenko, I. Nano-Objects and Ions in Liquid Crystals: Ion Trapping Effect and Related Phenomena. *Crystals* **2015**, *5*, 501–533. [CrossRef]
3. Urbanski, M. On the impact of nanoparticle doping on the electro-optic response of nematic hosts. *Liq. Cryst. Today* **2015**, *24*, 102–115. [CrossRef]
4. Blanc, C.; Coursault, D.; Lacaze, E. Ordering nano- and microparticles assemblies with liquid crystals. *Liq. Cryst. Rev.* **2013**, *1*, 83–109. [CrossRef]
5. Stamatoiu, O.; Mirzaei, J.; Feng, X.; Hegmann, T. Nanoparticles in liquid crystals and liquid crystalline nanoparticles. *Top. Curr. Chem.* **2012**, *318*, 331–394. [PubMed]
6. Brochard, F.; de Gennes, P.G. Theory of magnetic suspensions in liquid crystals. *J. Phys. (Paris)* **1970**, *31*, 691–708. [CrossRef]
7. Mertelj, A.; Lisjak, D. Ferromagnetic nematic liquid crystals. *Liq. Cryst. Rev.* **2017**, *5*, 1–33. [CrossRef]
8. Garbovskiy, Y.; Glushchenko, A. Liquid crystalline colloids of nanoparticles: Preparation, properties, and applications. *Solid State Phys.* **2011**, *62*, 1–74.
9. Garbovskiy, Y.; Zribi, O.; Glushchenko, A. Emerging Applications of Ferroelectric Nanoparticles in Materials Technologies, Biology and Medicine. In *Advances in Ferroelectrics*; Peláiz-Barranco, A., Ed.; InTech: Rijeka, Croatia, 2012; ISBN 978-953-51-0885-6. [CrossRef]
10. Mirzaei, J.; Reznikov, M.; Hegmann, T. Quantum dots as liquid crystal dopants. *J. Mater. Chem.* **2012**, *22*, 22350–22365. [CrossRef]
11. Klimusheva, G.; Mirnaya, T.; Garbovskiy, Y. Versatile Nonlinear-Optical Materials Based on Mesomorphic Metal Alkanoates: Design, Properties, and Applications. *Liq. Cryst. Rev.* **2015**, *3*, 28–57. [CrossRef]
12. Kumar, S. Discotic liquid crystal-nanoparticle hybrid systems. *NPG Asia Mater.* **2014**, *6*, e82. [CrossRef]
13. Rahman, M.; Lee, W. Scientific duo of carbon nanotubes and nematic liquid crystals. *J. Phys. D Appl. Phys.* **2009**, *42*, 063001. [CrossRef]
14. Yadav, S.P.; Singh, S. Carbon nanotube dispersion in nematic liquid crystals: An overview. *Prog. Mater. Sci.* **2016**, *80*, 38–76. [CrossRef]
15. Reznikov, Y.; Buchnev, O.; Tereshchenko, O.; Reshetnyak, V.; Glushchenko, A.; West, J. Ferroelectric nematic suspension. *Appl. Phys. Lett.* **2003**, *82*, 1917–1919. [CrossRef]
16. Ouskova, E.; Buchnev, O.; Reshetnyak, V.; Reznikov, Y.; Kresse, H. Dielectric relaxation spectroscopy of a nematic liquid crystal doped with ferroelectric $Sn_2P_2S_6$ nanoparticles. *Liq. Cryst.* **2003**, *30*, 1235–1239. [CrossRef]
17. Buchnev, O.; Glushchenko, A.; Reznikov, Y.; Reshetnyak, V.; Tereshchenko, O.; West, J. Diluted ferroelectric suspension of $Sn_2P_2S_6$ nanoparticles in nematic liquid crystal. In *Proceedings of the Ninth International Conference on Nonlinear Optics of Liquid and Photorefractive, Bellingham, WA, USA, 3 December 2003*; Volume 5257, pp. 7–12.

18. Buchnev, O.; Ouskova, E.; Reznikov, Y.; Reshetnyak, V.; Kresse, H.; Grabar, A. Enhanced dielectric response of liquid crystal ferroelectric suspension. *Mol. Cryst. Liq. Cryst.* **2004**, *422*, 47–55. [CrossRef]
19. Reshetnyak, V. Effective dielectric function of ferroelectric LC suspensions. *Mol. Cryst. Liq. Cryst.* **2004**, *421*, 219–224. [CrossRef]
20. Buchnev, O.; Cheon, C.I.; Glushchenko, A.; Reznikov, Y.; West, J.L. New non-synthetic method to modify properties of liquid crystals using micro- and nano-particles. *J. Soc. Inf. Disp.* **2005**, *13*, 749–754. [CrossRef]
21. Glushchenko, A.V. Ferroelectric particles in liquid crystals and their outcome. In Proceedings of the International Conference on Optics of Liquid Crystals, Sand Clear Water Beach, FL, USA, 1–7 October 2005.
22. Reznikov, Y.; Glushchenko, A.; Reshetnyak, V.; West, J. Liquid Crystal Cell Comprising Ferroelectric Particle Suspensions. Patent WO 2003060598 A2, 24 July 2003.
23. Reznikov, Y.; Buluy, O.; Tereshchenko, O.; Glushchenko, A.; West, J. Ferroelectric particles liquid crystal dispersions. *Proc. SPIE* **2005**, *5741*, 171.
24. Cheon, C.I.; Li, L.; Glushchenko, A.; West, J.L.; Reznikov, Y.; Kim, J.S.; Kim, D.H. Electro-optics of liquid crystals doped with ferroelectric nano-powder. *SID Tech. Digest* **2005**, *2*, 45. [CrossRef]
25. Li, F.; Buchnev, O.; Cheon, C.; Glushchenko, A.; Reshetnyak, V.; Reznikov, Y.; Sluckin, T.J.; West, J.L. Orientational coupling amplification in ferroelectric nematic colloids. *Phys. Rev. Lett.* **2006**, *97*, 147801. [CrossRef] [PubMed]
26. Glushchenko, A.; Cheon, C.; West, J.; Li, F.; Büyüktanir, E.; Reznikov, Y.; Buchnev, A. Ferroelectric particles in liquid crystals: Recent frontiers. *Mol. Cryst. Liq. Cryst.* **2006**, *453*, 227–237. [CrossRef]
27. Reshetnyak, V.Y.; Shelestiuk, S.M.; Sluckin, T.J. Freedericksz transition threshold in nematic liquid crystals filled with ferroelectric nano-particles. *Mol. Cryst. Liq. Cryst.* **2006**, *454*, 201/[603]–206/[608]. [CrossRef]
28. Li, F.; West, J.; Glushchenko, A.; Cheon, C.; Reznikov, Y. Ferroelectric nanoparticle/liquid-crystal colloids for display applications. *J. SID* **2006**, *14*, 523–527. [CrossRef]
29. Glushchenko, A.V.; West, J.L.; Li, F.; Zhang, K.; Atkuri, H.M. Electro-Optic Properties of Ferroelectric Nanoparticles in Liquid Crystal Dispersions. In Proceedings of the SID 2007 International Symposium, Long Beach, CA, USA, 20–25 May 2007.
30. Glushchenko, A.; Buchnev, O.; Iljin, A.; Kurochkin, O.; Reznikov, Y. Cholesteric colloid of ferroelectric nano-particles. In *SID Symposium Digest of Technical Papers*; Blackwell Publishing Ltd.: Hoboken, NJ, USA, 2007; Volume 38, pp. 1086–1089.
31. Glushchenko, A.; Cheon, C.; West, J.; Reznikov, Y. Invited paper: Applications of ferroelectric particles/liquid crystal colloids. *Proc. SPIE* **2007**, *6487*, 6487-0T.
32. Cook, G.; Glushchenko, A.V.; Reshetnyak, V.Y.; Saleh, M.A.; Evans, D.R. Hybrid Liquid Crystal Inorganic Photorefractives. In Proceedings of the 12th International Topical Meeting on Optics of Liquid Crystals (OLC-07), Puebla, Mexico, 1–5 October 2007.
33. Kaczmarek, M.; Dyadyusha, A.; D'Alessandro, G.; Buchnev, O. Hybrid liquid crystal nanomaterials with improved photorefractive response. In Proceedings of the Photorefractive Effects, Photosensitivity, Fiber Gratings, Photonic Materials and More, Squaw Creek, CA, USA, 14 October 2007.
34. West, J.L.; Li, F.; Zhang, K.; Atkuri, H.M.; Glushchenko, A.V. Electro-optic properties of ferroelectric nanoparticle/liquid crystal dispersions. In *SID Symposium Digest of Technical Papers*; Blackwell Publishing Ltd.: Hoboken, NJ, USA, 2007; Volume 38, pp. 1090–1092.
35. Buchnev, O.; Dyadyusha, A.; Kaczmarek, M.; Reshetnyak, V.; Reznikov, Y. Enhanced two-beam coupling in colloids of ferroelectric nanoparticles in liquid crystals. *J. Opt. Soc. Am. B* **2007**, *24*, 1512–1516. [CrossRef]
36. Čopič, M.; Mertelj, A.; Buchnev, O.; Reznikov, Y. Coupled director and polarization fluctuations in suspensions of ferroelectric nanoparticles in nematic liquid crystals. *Phys. Rev. E* **2007**, *76*, 011702. [CrossRef] [PubMed]
37. Cook, G.; Glushchenko, A.; Reshetnyak, V.; Beckel, E.; Saleh, M.; Evans, D. Liquid crystal inorganic hybrid photorefractives. In Proceedings of the 2008 IEEE/LEOS Winter Topical Meeting Series, Sorrento, Italy, 14–16 January 2008; pp. 129–130.
38. Cook, G.; Glushchenko, A.V.; Reshetnyak, V.; Griffith, A.T.; Saleh, M.A.; Evans, D.R. Nanoparticle doped organic-inorganic hybrid photorefractives. *Opt. Express* **2008**, *16*, 4015–4022. [CrossRef] [PubMed]
39. Kaczmarek, M.; Buchnev, O.; Nandhakumar, I. Ferroelectric nanoparticles in low refractive index liquid crystals for strong electro-optic response. *Appl. Phys. Lett.* **2008**, *92*, 103307. [CrossRef]

40. Atkuri, H.; Cook, G.; Evans, D.R.; Glushchenko, A.; Reshetnyak, V.; Reznikov, Y.; West, J.; Zhang, K. Preparation of ferroelectric nanoparticles and their use in organic-inorganic liquid crystal hybrid photorefractives. *J. Opt. A Pure Appl. Opt.* **2009**, *11*, 024006. [CrossRef]

41. Mikulko, A.; Arora, P.; Glushchenko, A.; Lapanik, A.; Haase, W. Complementary studies of BaTiO$_3$ nanoparticles suspended in a ferroelectric liquid-crystalline mixture. *Europhys. Lett.* **2009**, *87*, 27009. [CrossRef]

42. Kurochkin, O.; Buchnev, O.; Iljin, A.; Park, S.K.; Kwon, S.B.; Grabar, O.; Reznikov, Y. A colloid of ferroelectric nanoparticles in a cholesteric liquid crystal. *J. Opt. A Pure Appl. Opt.* **2009**, *11*, 024003. [CrossRef]

43. Atkuri, H.M.; Zhang, K.; West, J.L. Fabrication of paraelectric nanocolloidal liquid crystals. *Mol. Cryst. Liq. Cryst.* **2009**, *508*, 183/[545]–190/[552]. [CrossRef]

44. Akimoto, M.; Kundu, S.; Isomura, K.; Hirayama, I.; Kobayashi, S.; Takatoh, K. Improvement of electro-optical characteristics of liquid crystal display by nanoparticle-embedded alignment layers. *Mol. Cryst. Liq. Cryst.* **2009**, *508*, 1/[363]–13/[375]. [CrossRef]

45. Scolari, L.; Gauza, S.; Xianyu, H.; Zhai, L.; Eskildsen, L.; Tanggaard Alkeskjold, T.; Wu, S.-T.; Bjarklev, A. Frequency tunability of solid-core photonic crystal fibers filled with nanoparticle-doped liquid crystals. *Opt. Express* **2009**, *17*, 3754–3764. [CrossRef] [PubMed]

46. Evans, D.R.; Cook, G.; Saleh, M.A. Recent advances in photorefractive two-beam coupling. *Opt. Mater.* **2009**, *31*, 1059–1060. [CrossRef]

47. Lopatina, L.M.; Selinger, J.V. Theory of ferroelectric nanoparticles in nematic liquid crystals. *Phys. Rev. Lett.* **2009**, *102*, 197802. [CrossRef] [PubMed]

48. Kurochkin, O.; Atkuri, H.; Buchnev, O.; Glushchenko, A.; Grabar, O.; Karapinar, R.; Reshetnyak, V.; West, J.; Reznikov, Y. Nano-colloids of Sn2P2S6 in nematic liquid crystal pentyl-cianobiphenile. *Condens. Matter Phys.* **2010**, *13*, 33701. [CrossRef]

49. Cook, G.; Barnes, J.L.; Basun, S.A.; Evans, D.R.; Ziolo, R.F.; Ponce, A.; Reshetnyak, V.Y.; Glushchenko, A.; Banerjee, P.P. Harvesting single ferroelectric domain stressed nanoparticles for optical and ferroic applications. *J. Appl. Phys.* **2010**, *108*, 064309. [CrossRef]

50. Cook, G.; Reshetnyak, V.Y.; Ponce, A.; Ziolo, R.F.; Basun, S.A.; Evans, D.R. Improved holographic beam coupling through selective harvesting of single domain ferroelectric nanoparticles. In Proceedings of the Biomedical Optics and 3-D Imaging, Miami, FL, USA, 12–14 April 2010.

51. Pereira, M.S.S.; Canabarro, A.A.; de Oliveira, I.N.; Lyra, M.L.; Mirantsev, L.V. A molecular dynamics study of ferroelectric nanoparticles immersed in a nematic liquid crystal. *Eur. Phys. J. E* **2010**, *31*, 81–87. [CrossRef] [PubMed]

52. Blach, J.-F.; Saitzek, S.; Legrand, C.; Dupont, L.; Henninot, J.-F.; Warenghem, M. BaTiO$_3$ ferroelectric nanoparticles dispersed in 5CB nematic liquid crystal: Synthesis and electro-optical characterization. *J. Appl. Phys.* **2010**, *107*, 074102. [CrossRef]

53. Gupta, M.; Satpathy, I.; Roy, A.; Pratibha, R. Nanoparticle induced director distortion and disorder in liquid crystal-nanoparticle dispersions. *J. Colloid Interface Sci.* **2010**, *352*, 292–298. [CrossRef] [PubMed]

54. Domenici, V.; Zupancic, B.; Laguta, V.V.; Belous, A.G.; V'yunov, O.I.; Remskar, M.; Zalar, B. PbTiO$_3$ nanoparticles embedded in a liquid crystalline elastomer matrix: Structural and ordering properties. *J. Phys. Chem. C* **2010**, *114*, 10782–10789. [CrossRef]

55. Liang, H.-H.; Xiao, Y.-Z.; Hsh, F.-J.; Wu, C.-C.; Lee, J.-Y. Enhancing the electro-optical properties of ferroelectric liquid crystals by doping ferroelectric nanoparticles. *Liq. Cryst.* **2010**, *37*, 255–261. [CrossRef]

56. Herrington, M.R.; Buchnev, O.; Kaczmarek, M.; Nandhakumar, I. The effect of the size of BaTiO$_3$ nanoparticles on the electro-optic properties of nematic liquid crystals. *Mol. Cryst. Liq. Cryst.* **2010**, *527*, 72/[228]–79/[235]. [CrossRef]

57. Cook, G.; Reshetnyak, V.Y.; Ziolo, R.F.; Basun, S.A.; Banerjee, P.P.; Evans, D.R. Asymmetric Freedericksz transitions from symmetric liquid crystal cells doped with harvested ferroelectric nanoparticles. *Opt. Express* **2010**, *18*, 17339–17345. [CrossRef] [PubMed]

58. Meneses-Franco, A.; Trujillo-Rojo, V.H.; Soto-Bustamante, E.A. Synthesis and characterization of pyroelectric nanocomposite formed of BaTiO$_3$ nanoparticles and a smectic liquid crystal matrix. *Phase Transit.* **2010**, *83*, 1037–1047. [CrossRef]

59. West, J.L.; Cheon, C., II; Glushchenko, A.V.; Reznikov, Y.; Li, F. Non-Synthetic Method for Modifying Properties of Liquid Crystals. U.S. Patent 7758773 B2, 20 July 2010.

60. Evans, D.R.; Basun, S.A.; Cook, G.; Pinkevych, I.P.; Reshetnyak, V.Y. Electric field interactions and aggregation dynamics of ferroelectric nanoparticles in isotropic fluid suspensions. *Phys. Rev. B* **2011**, *84*, 174111. [CrossRef]

61. Coondoo, I.; Goel, P.; Malik, A.; Biradar, A.M. Dielectric and polarization properties of BaTiO$_3$ nanoparticle/ferroelectric liquid crystal colloidal suspension. *Integr. Ferroelectr.* **2011**, *125*, 81–88. [CrossRef]

62. Paul, S.N.; Dhar, R.; Verma, R.; Sharma, S.; Dabrowski, R. Change in dielectric and electro-optical properties of a nematic material (6CHBT) due to the dispersion of BaTiO$_3$ nanoparticles. *Mol. Cryst. Liq. Cryst.* **2011**, *545*, 105/[1329]–111/[1335]. [CrossRef]

63. Basun, S.A.; Cook, G.; Reshetnyak, V.Y.; Glushchenko, A.V.; Evans, D.R. Dipole moment and spontaneous polarization of ferroelectric nanoparticles in a nonpolar fluid suspension. *Phys. Rev. B* **2011**, *84*, 024105. [CrossRef]

64. Shelestiuk, S.M.; Reshetnyak, V.Y.; Sluckin, T.J. Frederiks transition in ferroelectric liquid-crystal nanosuspensions. *Phys. Rev. E* **2011**, *83*, 041705. [CrossRef] [PubMed]

65. Lopatina, L.M.; Selinger, J.V. Maier-Saupe-type theory of ferroelectric nanoparticles in nematic liquid crystals. *Phys. Rev. E* **2011**, *84*, 041703. [CrossRef] [PubMed]

66. Ghosh, S.; Roy, S.K.; Acharya, S.; Chakrabarti, P.K.; Zurowska, M.; Dabrowski, R. Effect of multiferroic BiFeO$_3$ nanoparticles on electro-optical and dielectric properties of a partially fluorinated orthoconic antiferroelectric liquid crystal mixture. *EPL* **2011**, *96*, 47003. [CrossRef]

67. Wang, L.; He, W.; Xiao, X.; Wang, M.; Wang, M.; Yang, P.; Zhou, Z.; Yang, H.; Yu, H.; Lu, Y. Low voltage and hysteresis-free blue phase liquid crystal dispersed by ferroelectric nanoparticles. *J. Mater. Chem.* **2012**, *22*, 19629–19633. [CrossRef]

68. Evans, D.R.; Cook, G. Enhanced Dynamic Holography in Organic-Inorganic Hybrid Devices. In Proceedings of the Biomedical Optics and 3D Imaging, Miami, FL, USA, 28 April–2 May 2012.

69. Ghandevosyan, A.A.; Hakobyan, R.S. Decrease in the Threshold of Electric Freedericksz Transition in Nematic Liquid Crystals Doped with Ferroelectric Nanoparticles. *J. Contemp. Phys. (Armen. Acad. Sci.)* **2012**, *47*, 33–35. [CrossRef]

70. Hasegawa, M. Characterization of BaTiO$_3$ nanoparticles suspension in liquid crystals. *J. Photopolym. Sci. Technol.* **2012**, *25*, 295–299. [CrossRef]

71. Mertelj, A.; Cmok, L.; Copic, M.; Cook, G.; Evans, D.R. Critical behavior of director fluctuations in suspensions of ferroelectric nanoparticles in liquid crystals at the nematic to smectic-A phase transition. *Phys. Rev. E* **2012**, *85*, 021705. [CrossRef] [PubMed]

72. Lorenz, A.; Zimmermann, N.; Kumar, S.; Evans, D.R.; Cook, G.; Kitzerow, H.-S. Doping the nematic liquid crystal 5CB with milled BaTiO$_3$ nanoparticles. *Phys. Rev. E* **2012**, *86*, 051704. [CrossRef] [PubMed]

73. Warenghem, M.; Henninot, J.-F.; Blach, J.F.; Buchnev, O.; Kaczmarek, M. Combined ellipsometry and refractometry technique for characterisation of liquid crystal based nanocomposites. *Rev. Sci. Instrum.* **2012**, *83*, 035103. [CrossRef] [PubMed]

74. Manna, S.K.; Sinha, A. Development of a Phenomenological Model on Surface Stabilized Ferroelectric Liquid Crystal Nanocomposite. *IOSR J. Appl. Phys. (IOSRJAP)* **2012**, *1*, 33–38. [CrossRef]

75. Klein, S.; Richardson, R.M.; Greasty, R.; Jenkins, R.; Stone, J.; Thomas, M.R.; Sarua, A. The influence of suspended nanoparticles on the Frederiks threshold of the nematic host. *Philos. Trans. R. Soc. A* **2013**, *371*, 20120253. [CrossRef] [PubMed]

76. Lorenz, A.; Zimmermann, N.; Kumar, S.; Evans, D.R.; Cook, G.; Martínez, M.F.; Kitzerow, H.-S. X-ray scattering of nematic liquid crystal nanodispersion with negative dielectric anisotropy. *Appl. Opt.* **2013**, *52*, E1–E5. [CrossRef] [PubMed]

77. Rudzki, A.; Evans, D.R.; Cook, G.; Haase, W. Size dependence of harvested BaTiO$_3$ nanoparticles on the electro-optic and dielectric properties of ferroelectric liquid crystal nanocolloids. *Appl. Opt.* **2013**, *52*, E6–E14. [CrossRef] [PubMed]

78. Ganguly, P.; Kumar, A.; Tripathi, S.; Haranath, D.; Biradar, A.M. Faster and highly luminescent ferroelectric liquid crystal doped with ferroelectric BaTiO$_3$ nanoparticles. *Appl. Phys. Lett.* **2013**, *102*, 222902. [CrossRef]

79. Lahiri, T.; Majumder, T.P.; Ghosh, N.K. Theory of nanoparticles doped in ferroelectric liquid crystals. *J. Appl. Phys.* **2013**, *113*, 064308. [CrossRef]

80. Sigdel, K.P.; Iannacchione, G.S. Calorimetric study of phase transitions in ocylcyanobiphenyl-barium titanate nanoparticle dispersions. *J. Chem. Phys.* **2013**, *139*, 204906. [CrossRef] [PubMed]

81. Chaudharya, A.; Malik, P.; Mehra, R.; Raina, K.K. Influence of ZnO nanoparticle concentration on electro-optic and dielectric properties of ferroelectric liquid crystal mixture. *J. Mol. Liq.* **2013**, *188*, 230–236. [CrossRef]

82. Lorenz, A.; Zimmermann, N.; Kumar, S.; Evans, D.R.; Cook, G.; Martínez, M.F.; Kitzerow, H.-S. Doping a Mixture of Two Smectogenic Liquid Crystals with Barium Titanate Nanoparticles. *J. Phys. Chem. B* **2013**, *117*, 937–941. [CrossRef] [PubMed]

83. Lorenz, A.; AgraKooijman, D.M.; Zimmermann, N.; Kitzerow, H.-S.; Evans, D.R.; Kumar, S. Bilayers in nanoparticle-doped polar mesogens. *Phys. Rev. E* **2013**, *88*, 062505. [CrossRef] [PubMed]

84. Hakobyan, M.R.; Alaverdyan, R.B.; Hakobyan, R.S.; Chilingaryan, Y.S. Enhanced physical properties of nematics doped with ferroelectric nanoparticles. *Armen. J. Phys.* **2014**, *7*, 11–18.

85. Shukla, R.K.; Liebig, C.M.; Evans, D.R.; Haase, W. Electro-optical behaviour and dielectric dynamics of harvested ferroelectric $LiNbO_3$ nanoparticle-doped ferroelectric liquid crystal nanocolloids. *RSC Adv.* **2014**, *4*, 18529–18536. [CrossRef]

86. Basu, R.; Garvey, A. Effects of ferroelectric nanoparticles on ion transport in a liquid crystal. *Appl. Phys. Lett.* **2014**, *105*, 151905. [CrossRef]

87. Hakobyan, M.R. Onsager theory of nematic liquid crystals doped with ferroelectric nanoparticles. *Proc. Yerevan State Univ.* **2014**, *2*, 54–59.

88. Darla, M.R.; Hegde, S.; Varghese, S. Effect of $BaTiO_3$ Nanoparticle on Electro-Optical Properties of Polymer Dispersed Liquid Crystal Displays. *J. Cryst. Process Technol.* **2014**, *4*, 60–63. [CrossRef]

89. Basu, R. Soft memory in a ferroelectric nanoparticle-doped liquid crystal. *Phys. Rev. E* **2014**, *89*, 022508. [CrossRef] [PubMed]

90. Podoliak, N.; Buchnev, O.; Herrington, M.; Mavrona, E.; Kaczmarek, M.; Kanaras, A.G.; Stratakis, E.; Blach, J.; Henninot, J.; Warenghem, M. Elastic constants, viscosity and response time in nematic liquid crystals doped with ferroelectric nanoparticles. *RSC Adv.* **2014**, *4*, 46068–46074. [CrossRef]

91. Mavrona, E.; Chodorow, U.; Barnes, M.E.; Parka, J.; Palka, N.; Saitzek, S.; Blach, J.-F.; Apostolopoulos, V.; Kaczmarek, M. Refractive indices and birefringence of hybrid liquid crystal-nanoparticles composite materials in the terahertz region. *AIP Adv.* **2015**, *5*, 077143. [CrossRef]

92. Kurochkin, O.; Mavrona, E.; Apostolopoulos, V.; Blach, J.-F.; Henninot, J.-F.; Kaczmarek, M.; Saitzek, S.; Sokolova, M.; Reznikov, Y. Electrically charged dispersions of ferroelectric nanoparticles. *Appl. Phys. Lett.* **2015**, *106*, 043111. [CrossRef]

93. Garbovskiy, Y.; Glushchenko, I. Ion trapping by means of ferroelectric nanoparticles, and the quantification of this process in liquid crystals. *Appl. Phys. Lett.* **2015**, *107*, 041106. [CrossRef]

94. Lin, Y.; Douali, R.; Dubois, F.; Segovia-Mera, A.; Daoudi, A. On the phase transitions of $8CB/Sn_2P_2S_6$ liquid crystal nanocolloids. *Eur. Phys. J. E* **2015**, *38*, 103. [CrossRef] [PubMed]

95. Dalir, N.; Javadian, S.; Gilani, A.G. The ferroelectricity effect of nanoparticles on thermodynamics and electro-optics of novel cyanobiphenyl eutectic binary mixture liquid crystals. *J. Mol. Liq.* **2015**, *209*, 336–345. [CrossRef]

96. Rasna, M.V.; Cmok, L.; Evans, D.R.; Mertelj, A.; Dhara, S. Phase transitions, optical, dielectric and viscoelastic properties of colloidal suspensions of $BaTiO_3$ nanoparticles and cyanobiphenyl liquid crystals. *Liq. Cryst.* **2015**, *42*, 1059–1067. [CrossRef]

97. Xu, X.W.; Zhang, X.W.; Luo, D.; Dai, H.T. Low voltage polymer-stabilized blue phase liquid crystal reflective display by doping ferroelectric nanoparticles. *Opt. Express* **2015**, *23*, 32267–32273. [CrossRef] [PubMed]

98. Cîrtoaje, C.; Petrescun, E.; Stoian, V. Electrical Freedericksz transitions in nematic liquid crystals containing ferroelectric nanoparticles. *Physica E* **2015**, *67*, 23–27. [CrossRef]

99. Nayek, P.; Li, G. Superior electro-optic response in multiferroic bismuth ferrite nanoparticle doped nematic liquid crystal device. *Sci. Rep.* **2015**, *5*, 10845. [CrossRef] [PubMed]

100. Kumar, P.; Kishore, A.; Sinha, A. Analog switching in the nanocolloids of ferroelectric liquid crystals. *Appl. Phys. Lett.* **2016**, *108*, 262903. [CrossRef]

101. Evans, D.R.; Cook, G.; Reshetnyak, V.Y.; Liebig, C.M.; Basun, S.A.; Banerjee, P.P. Inorganic-Organic Photorefractive Hybrids. In *Photorefractive Organic Materials and Applications*; Blanche, P.-A., Ed.; Springer: Berlin, Germany, 2016; Volume 240, pp. 223–247.

102. Hsiao, Y.-C.; Huang, S.-M.; Yeh, E.-R.; Lee, W. Temperature-dependent electrical and dielectric properties of nematic liquid crystals doped with ferroelectric particles. *Displays* **2016**, *44*, 61–65. [CrossRef]

103. Garbovskiy, Y. Electrical properties of liquid crystal nanocolloids analysed from perspectives of the ionic purity of nano-dopants. *Liq. Cryst.* **2016**, *43*, 648–653. [CrossRef]

104. Mukherjee, P.K. Effect of ferroelectric nanoparticles on the isotropic-smectic-A phase transition. *EPL* **2016**, *114*, 56002. [CrossRef]

105. Shukla, R.K.; Evans, D.R.; Haase, W. Ferroelectric BaTiO$_3$ and LiNbO$_3$ nanoparticles dispersed in ferroelectric liquid crystal mixtures: Electrooptic and dielectric parameters influenced by properties of the host, the dopant and the measuring cell. *Ferroelectrics* **2016**, *500*, 141–152. [CrossRef]

106. Ibragimov, T.D.; Imamaliyev, A.R.; Bayramov, G.M. Formation of local electric fields in the ferroelectric BaTiO$_3$ particles-liquid crystal colloids. *Ferroelectrics* **2016**, *495*, 60–68. [CrossRef]

107. Mishra, M.; Dabrowski, R.S.; Dhar, R. Thermodynamical, optical, electrical and electro-optical studies of a room temperature nematic liquid crystal 4-pentyl-4'-cyanobiphenyl dispersed with barium titanate nanoparticles. *J. Mol. Liq.* **2016**, *213*, 247–254. [CrossRef]

108. Ibragimov, T.D.; Imamaliyev, A.R.; Bayramov, G.M. Electro-optic properties of the BaTiO$_3$—Liquid crystal 5CB colloid. *Optik* **2016**, *127*, 2278–2281. [CrossRef]

109. Ibragimov, T.D.; Imamaliyev, A.R.; Bayramov, G.M. Influence of barium titanate particles on electro-optic characteristicsof liquid crystalline mixture H-37. *Optik* **2016**, *127*, 1217–1220. [CrossRef]

110. Shim, H.; Lyu, H.-K.; Allabergenov, B.; Garbovskiy, Y.; Glushchenko, A.; Choi, B. Enhancement of frequency modulation response time for polymer-dispersed liquid crystal. *Liq. Cryst.* **2016**, *43*, 1390–1396. [CrossRef]

111. Shim, H.; Lyu, H.-K.; Allabergenov, B.; Garbovskiy, Y.; Glushchenko, A.; Choi, B. Switchable Response of Ferroelectric Nanoparticle Doped Polymer-Dispersed Liquid Crystals. *J. Nanosci. Nanotechnol.* **2016**, *16*, 11125–11129. [CrossRef]

112. Lin, Y.; Daoudi, A.; Segovia-Mera, A.; Dubois, F.; Legrand, C.; Douali, R. Electric field effects on phase transitions in the 8CB liquid crystal doped with ferroelectric nanoparticles. *Phys. Rev. E* **2016**, *93*, 062702.

113. Rzoska, S.J.; Starzonek, S.; Drozd-Rzoska, A.; Czuprynski, K.; Chmiel, K.; Gaura, G.; Michulec, A.; Szczypek, B.; Walas, W. Impact of BaTiO$_3$ nanoparticles on pretransitional effects in liquid crystalline dodecylcyanobiphenyl. *Phys. Rev. E* **2016**, *93*, 020701(R). [CrossRef] [PubMed]

114. Kovalchuk, O.V.; Kovalchuk, T.M.; Kucheriavchenkova, N.M.; Sydorchuk, V.V.; Khalameida, S.V. Multiferroic based on nematic liquid crystal and nanoparticles. *Semicond. Phys. Quantum Electron. Optoelectron.* **2016**, *19*, 285–289. [CrossRef]

115. Al-Zangana, S.; Turner, M.; Dierking, I. A comparison between size dependent paraelectric and ferroelectric BaTiO$_3$ nanoparticle doped nematic and ferroelectric liquid crystals. *J. Appl. Phys.* **2017**, *121*, 085105. [CrossRef]

116. Dubey, R.; Mishra, A.; Singh, K.N.; Alapati, P.R.; Dhar, R. Electric behaviour of a Schiff's base liquid crystal compound doped with a low concentration of BaTiO$_3$ nanoparticles. *J. Mol. Liq.* **2017**, *225*, 496–501. [CrossRef]

117. Mukherjee, P.K. Effect of ferroelectric nanoparticles on the dielectric permittivity in the isotropic phase of the isotropic-smectic-A phase transition. *J. Mol. Liq.* **2017**, *225*, 462–466. [CrossRef]

118. Lin, Y.; Daoudi, A.; Dubois, F.; Segovia-Mera, A.; Legrand, C.; Douali, R. Correlation between dielectric properties and phase transitions of 8CB/Sn$_2$P$_2$S$_6$ liquid crystal nanocolloids. *J. Mol. Liq.* **2017**, *232*, 123–129. [CrossRef]

119. Poursamad, J.B.; Hallaji, T. Freedericksz transition in smectic-A liquid crystals doped by ferroelectric nanoparticles. *Physica B* **2017**, *504*, 112–115. [CrossRef]

120. Chodorow, U.; Mavrona, E.; Palka, N.; Strzezysz, O.; Garbat, K.; Saitzek, S.; Blach, J.F.; Apostolopoulos, V.; Kaczmarek, M.; Parka, J. Terahertz properties of liquid crystals doped with ferroelectric BaTiO$_3$ nanoparticles. *Liq. Cryst.* **2017**, *44*, 1207–1215. [CrossRef]

121. Garbovskiy, Y. Nanoparticle enabled thermal control of ions in liquid crystals. *Liq. Cryst.* **2017**, *44*, 948–955. [CrossRef]

122. Kumar, R.; Srikanth, K.; Nandiraju, T.; Rao, V.S.; Ghosh, S. Elastic and dielectric properties of ferroelectric nanoparticles/bent-core nematic liquid crystal blend. *Eur. Phys. J. E* **2017**, *40*, 75.

123. Lin, Y.; Daoudi, A.; Dubois, F.; Blacj, J.F.; Henninot, J.F.; Kurochkin, O.; Grabar, A.; Segoiva-Mera, A.; Legrand, C.; Douali, R. A comparative study of nematic liquid crystals doped with harvested and nonm-harvested ferroelectric nanoparticles: Phase tranasitions and dielectric properties. *RSC Adv.* **2017**, *7*, 35438–35444. [CrossRef]

124. Garbovskiy, Y.; Glushchenko, A. Optical/ferroelectric characterization of $BaTiO_3$ and $PbTiO_3$ colloidal nanoparticles and their applications in hybrid materials technologies. *Appl. Opt.* **2013**, *52*, E34–E39. [CrossRef] [PubMed]

125. Reznikov, Y.; Glushchenko, A.; Garbovskiy, Y. Ferromagnetic and Ferroelectric Nanoparticles in Liquid Crystals. In *Liquid Crystals with Nano and Microparticles*; World Scientific Publishing Co. Pte. Ltd.: Singapore, 2016; Volume 2, pp. 657–693. ISBN 978-981-4619-25-7.

126. Matsuyama, A. Phase separations in mixtures of a liquid crystal and a nanocolloidal particle. *J. Chem. Phys.* **2009**, *131*, 204904. [CrossRef] [PubMed]

127. Gruverman, A.; Kholkin, A. Nanoscale ferroelectrics: Processing, characterization and future trends. *Rep. Prog. Phys.* **2006**, *69*, 2443–2474. [CrossRef]

128. Prodanov, M.F.; Pogorelova, N.V.; Kryshtal, A.P.; Klymchenko, A.S.; Mely, Y.; Semynozhenko, V.P.; Krivoshey, A.I.; Reznikov, Y.A.; Yarmolenko, S.N.; Goodby, J.W.; et al. Thermodynamically Stable Dispersions of Quantum Dots in a Nematic Liquid Crystal. *Langmuir* **2013**, *29*, 9301–9309. [CrossRef] [PubMed]

129. Zribi, O.; Garbovskiy, Y.; Glushchenko, A. Single step colloidal processing of stable aqueous dispersions of ferroelectric nanoparticles for biomedical imaging. *Mater. Res. Express* **2014**, *1*, 045401. [CrossRef]

130. Garbovskiy, Y. Switching between purification and contamination regimes governed by the ionic purity of nanoparticles dispersed in liquid crystals. *Appl. Phys. Lett.* **2016**, *108*, 121104. [CrossRef]

131. Garbovskiy, Y. Ions in liquid crystals doped with nanoparticles: Conventional and counterintuitive temperature effects. *Liq. Cryst.* **2017**, *44*, 1402–1408. [CrossRef]

nanomaterials

MDPI

Article

Synthesis of Distinct Iron Oxide Nanomaterial Shapes Using Lyotropic Liquid Crystal Solvents

Seyyed Muhammad Salili [1,†], **Matthew Worden** [2,†], **Ahlam Nemati** [1], **Donald W. Miller** [3] **and Torsten Hegmann** [1,2,*]

1   Chemical Physics Interdisciplinary Program, Liquid Crystal Institute, Kent State University, Kent, OH 44242-0001, USA; ssalili@kent.edu (S.M.S.); anemati@kent.edu (A.N.)
2   Department of Chemistry and Biochemistry, Kent State University, Kent, OH 44242-0001, USA; mworden@austin.utexas.edu
3   Department of Pharmacology and Therapeutics, University of Manitoba, Winnipeg, MB R3E 0T6, Canada; Donald.Miller@umanitoba.ca
*   Correspondence: thegmann@kent.edu; Tel.: +1-330-672-7770
†   These authors contributed equally to this work.

Received: 29 June 2017; Accepted: 30 July 2017; Published: 2 August 2017

Abstract: A room temperature reduction-hydrolysis of Fe(III) precursors such as $FeCl_3$ or $Fe(acac)_3$ in various lyotropic liquid crystal phases (lamellar, hexagonal columnar, or micellar) formed by a range of ionic or neutral surfactants in H_2O is shown to be an effective and mild approach for the preparation of iron oxide (IO) nanomaterials with several morphologies (shapes and dimensions), such as extended thin nanosheets with lateral dimensions of several hundred nanometers as well as smaller nanoflakes and nanodiscs in the tens of nanometers size regime. We will discuss the role of the used surfactants and lyotropic liquid crystal phases as well as the shape and size differences depending upon when and how the resulting nanomaterials were isolated from the reaction mixture. The presented synthetic methodology using lyotropic liquid crystal solvents should be widely applicable to several other transition metal oxides for which the described reduction-hydrolysis reaction sequence is a suitable pathway to obtain nanoscale particles.

Keywords: iron oxide nanoparticles; magnetic nanoparticles; nanosheets; nanodiscs; nanoflakes; lyotropic liquid crystals; template syntheses

1. Introduction

The effects of surface chemistry and other properties of functionalized iron oxide nanoparticles (IO NPs) on cell uptake are fairly well established. This is best seen in the large number of reviews published on this topic over the past decade [1–27]. Especially multifunctional IO NPs (including surface chemistries for targeting, imaging, and drug delivery, among others) have attracted enormous attention. An often-overlooked aspect, however, is the effect that particle morphology (shape) may have on the theranostic (therapeutic + diagnostic) properties of IO NPs. This is unsurprising, as the vast majority of publications deal with very similar core shapes and sizes. The most common synthetic methods for creating functional IO NPs (i.e., the co-precipitation method by Massart [28], and the solvothermal method by Sun [29]) yield quasi-spherical particles or IO cubes in the range of tens of nanometers or smaller. While there are numerous papers that describe methods for creating non-spherical IO nanostructures [30] (particularly wires [23,31], rods [32–36], cubes [37–39], and sheets [40]), fewer of those are used or investigated for in-depth (bio)medical research. While there is a wealth of data for T2 MRI contrast enhancement and magnetic hyperthermia use of such IO nanoshapes [41], specific cell uptake [42] in various relevant tissue or cancer cell lines as well as their intracellular behavior have remained, to a large extent, unexplored.

1.1. Shape Effects on Particle Translocation and Endocytosis

Vacha and co-workers conducted molecular dynamics simulations that investigated the effect shape may have on the speed and ease with which nanoparticles may undergo endocytosis [43]. The researchers simulated a model set of nanoparticles, including perfectly spherical, cylindrical, and "spherocylinders" (i.e., cylinders with rounded ends) made of theoretical hydrophilic components they term "beads". Their computational model compares how spherocylindrical particles and spherical particles undergo endocytosis. The model clearly demonstrates that, with all other surface properties equal, the spherocylindrical particle will pass through the cell membrane much more quickly and efficiently than simple spherical particles with a similar size. The researchers explain these findings by citing Helfrich theory of membrane elasticity, which states that the ability of a particle to be encapsulated by a membrane is inversely proportional to its mean curvature. So, while the attractive energy per unit area is the same for each particle (since both have the same percentage of surface ligands), the mean curvature of the spherocylinder is half that of the sphere and it can thus more efficiently pass through the cell membrane.

Nangia et al. modeled particles of various shapes ranging from spheres and cubes to cone and rice-like geometries in order to investigate translocation rates through cell membranes [44]. In these simulations, the particles were composed of gold. The interactions between particles and the modeled cell membranes were due to shells of charge (both positive and negative, depending on the model) on the surface of the particles, rather than any intrinsic properties of gold. As such, the actual particle composition did not affect the results. Unsurprisingly the researchers found that, regardless of particle shape, negative surface charges on the particles inhibited translocation, while translocation rates increased with increasing positive surface charge. When these surface properties were accounted for, the effect of shape on translocation was clear and dramatic. For the maximum assigned charge density, they found that spherical particles had a free energy barrier for translocation of approximately 60 kJ/mol, giving the second lowest barrier and thus the second highest translocation rate. While these results were superior to those calculated for most of the other shapes, the rice-like particles were found to interact and pass through the cell membrane instantaneously—that is, the free energy barrier was effectively zero. The researchers explain that this is due to the anisotropic shape of the rice-like particles.

Moving away from computational models, Zhang et al. conducted hands-on experiments in which they compared cell uptake of polystyrene nanospheres with two-dimensional "nanodiscs" [45]. They synthesized both kinds of particles with diameters on the long axis of roughly 20 nm, but the discs were constrained to 2–3 nm along the perpendicular axis. The particles were then compared in cell permeation studies on HeLa cells (human cervical cancer cell line). The researchers found that the nanodiscs preferentially associated with the cell membrane, rarely passing into the cell endoplasm. This was in stark contrast to the nanospheres, which entered into the cells without accumulating along the membrane. They concluded from this that the nanodiscs were retained on the membrane at an 8-fold higher ratio as compared with the nanospheres. These data suggest that these differences are due to the fact that the 2-dimensionality of the nanodiscs allows them to enter into and between the phospholipid bilayers that compose the cell membrane. The nanospheres, which are too large to be maintained between the bilayers, disrupt the membrane to a much greater degree, leading to endocytosis.

1.2. The Influence of Iron Oxide Nanoparticle Shape

These earlier results on polystyrene nanoshapes are in stark contrast to our own data on IO nanospheres and nanobricks [42]. We examined the influence of IO NP shape in regulating preferential uptake specifically in endothelial cells (brain, lung). NPs targeting endothelial cells to treat diseases such as cancer, oxidative stress, and inflammation have traditionally relied on ligand-receptor based delivery. Spherical (diameter: 10 nm) and brick-shaped IO NPs (polyhedral bricks with 10 nm $\times$ 15 nm $\times$ 5 nm; $w \times d \times h$) were synthesized with identical negatively charged surface

EDTS coating (EDTS = *N*-(trimethoxysilylpropyl) ethylenediaminetriacetate trisodium salt; with a ζ-potential around −40 mV). These nanobricks were synthesized using Triton X surfactants in the hexagonal columnar or lamellar lyotropic liquid crystal (LLC) phase (Figure 1A) [46]. The nanobricks showed a significantly greater uptake profile in endothelial cells compared to the IO nanospheres. Furthermore, application of an external magnetic field significantly enhanced the uptake of the nanobricks but not the nanospheres. Transmission electron microscopy (TEM) revealed differential internalization of nanobricks in endothelial cells compared to epithelial cells. Given the reduced uptake of nanobricks in endothelial cells treated with caveolin inhibitors, the increased expression of caveolin-1 in endothelial cells compared to epithelial cells, and the ability of IO NP nanobricks to interfere with caveolae-mediated endocytosis process, a caveolae-mediated pathway was proposed as the working mechanism for the differential internalization of nanobricks in endothelial cells [42].

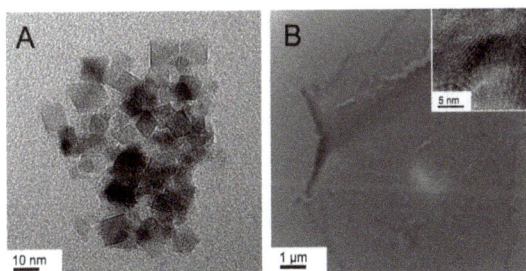

Figure 1. Transmission electron microscopy (TEM) images of previously synthesized IO nanostructures: (**A**) polyhedral nanobricks prepared by co-precipitation of Fe(III)- and Fe(II)-precursors (2:1 ratio) in the LLC phases (lamellar and hexagonal columnar) of Triton X-100 or Triton X-45 in H_2O [46]. Using TEM tomography the height of the nanobricks was determined to be ~5 nm. Reproduced from Ref. [46] with permission from The Royal Society of Chemistry; (**B**) Large nanosheets (lateral dimensions up to several micron) prepared in the Col$_h$ LLC phase formed by Triton X-100 in water using the reduction/hydrolysis method feature heights of 5 or 10 nm (inset shows higher resolution) [47]. Reproduced from Ref. [47] with permission from The Royal Society of Chemistry.

The research briefly discussed above, while clearly intended as a small sampling rather than an exhaustive review, demonstrates that morphology affects how particles interact with cells in profound and important ways. It demonstrates how tuning the shape of particles intended for biomedical applications—IO NPs included—may help us enhance and select certain properties, such the rate of cell uptake or particle interactions with the cells.

A cornerstone of such biomedical studies, particularly for clinically relevant IO nanomaterials, is the development of synthetic methods that create non-spherical IO nanoparticles in a consistent fashion with both shape and size control. Our first synthetic approach made use of a simple, room temperature reduction-hydrolysis of $FeCl_3$ in H_2O/Triton X-100 serving as a lyotropic hexagonal columnar LC template resulting in highly crystalline iron/iron oxide nanosheets [47]. These nanosheets, however, satisfied the requirements for biomedical applications only in one physical dimension, the height of 5 or 10 nm (multiples of 5) as determined by TEM tomography, which was attributed to the coalescence of smaller nanocrystallites to the finally obtained nanosheets in the hexagonal LLC template of Triton X-100 in water with a lattice parameter $d_{10(hex)}$ of around 5–6 nm.

In contrast, the lateral dimensions of the as-obtained I/IO nanosheets were significantly larger, extending in some cases over several hundreds of nanometers (Figure 1B), which is why we tested these for supercapacitor rather than for biomedical applications, where cell uptake, in vivo transport in blood vessels, etc. require much smaller NP dimensions.

Tuning the reaction parameters, i.e., using Massart's co-precipitation of $FeCl_3$ and $FeCl_2$ (2:1 molar ratio) [28] in the LLC phases of Triton X-100 (hexagonal columnar phase) or Triton X-45 (lamellar phase) in water, rather than following our in-house developed reduction/hydrolysis pathway [48–51], resulted in the above-mentioned IO nanobricks with the desirable smaller overall dimensions suitable for endothelial cell uptake, magnetic hyperthermia, as well as MRI contrast enhancement studies. Nevertheless, we were intrigued by the possibility of using various LLC solvents to guide the formation of anisometric IO nanomaterials, and decided to continue to explore this reaction space.

We demonstrate here that both the type of LLC phase (including micellar solutions of some of the used surfactants in water) and the nature of the surfactant (ionic or neutral) play significant roles in the shape-selective syntheses of IO NPs. Following our reduction/hydrolysis pathway to IO nanomaterials, we also show that various shapes if IO NPs can be synthesized, isolated, and separated as the reactions proceeds. As discussed in our initial article describing this particular synthesis, the formation of H_2 gas during the reduction of the Fe(III) precursor using $NaBH_4$ leads to the formation of two fractions, a gel-like fraction that remains in the reaction vessel and a foam (froth) fraction that is collected separately (Figure 2A–C). We will show that both fractions usually contain dissimilar shapes and sizes of the final IO nanostructures, which are conveniently separated by the synthesis itself.

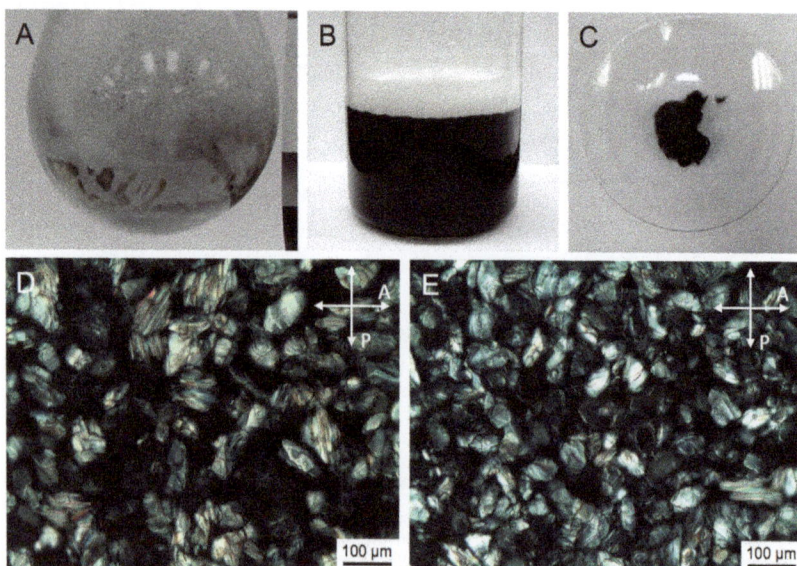

Figure 2. (**A**) Photograph showing the froth formation during after $NaBH_4$ addition; (**B**) Photograph showing the diluted froth solution from which as-synthesized nanostructures were isolated by magnetic separation and washing (see experimental section). For the complete experimental setup see photographs in the Supplementary Information; (**C**) Photograph showing the gel fraction that has undergone separate magnetic separation and washing; (**D,E**) Polarized light optical photomicrographs (crossed polarizer *P* and analyzer *A*) of the hexagonal columnar phase of one of the used surfactants (here CTAB) in water: (**D**) prior to the addition of the Fe(III)-precursor and (**E**) from a small sample taken during the reaction (The fact that no difference could be detected in the polarized optical microscopy (POM) textures prior to and after addition of the reducing agent indicated the persistence of the Col$_{hI}$ phase as the reaction proceeded).

2. Results

Since the originally published, large I/IO nanosheets were the result of a reduction/hydrolysis reaction in the hexagonal columnar phase formed by Triton X-100 in water, we were first interested in testing the reaction in the lamellar (Lα) phase. Since Triton-X-100 in water forms the Lα phase only below 10 °C at higher surfactant concentrations [52], we decided to substitute Triton X-100 with Triton X-45, which forms the Lα phase at room temperature and above over a quite large concentrations range (at least 20 to 60 wt %) [53].

In numerous instances, running the reaction at room temperature or below proved to be challenging because of the too high viscosities of the bulk LLC phases. For the syntheses with Triton X-45, all reactions were performed at 40 °C. We first varied the concentration of the $FeCl_3 \cdot 6H_2O$ (1 vs. 2 mmol), then the concentration of the surfactant (20 vs. 50 wt %), and finally the type of Fe(III) precursor (chloride vs. acetylacetonate or acac). The TEM images of the representative products are shown in Figure 3. A first, general observation regarding the obtained nanostructures is that the froth fractions (Figure 3A,C,E,G) consist of nanosheets with a varying amount of smaller nanoparticles in the size regime of 15 to 20 nm in diameter. The gel fractions, with the exception of the synthesis with the double amount of $FeCl_3$ (2 mmol vs. 1 mmol as in all other cases; Figure 3C,D), contained larger proportions of or even exclusively nanoparticles (Figure 3F,H).

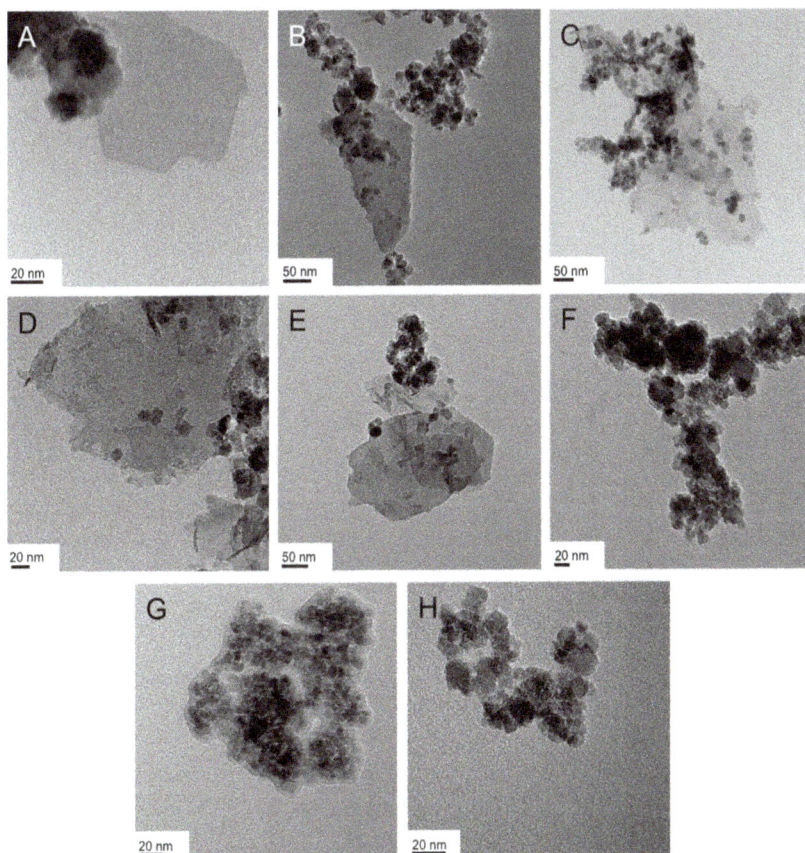

Figure 3. TEM images of the IO nanostructures obtained in the Lα phase of Triton X-45 in water: (**A,B**) 50 wt % Triton X-45, 1 mmol $FeCl_3$ ((**A**): froth, (**B**): gel); (**C,D**) 50 wt % Triton X-45, 2 mmol $FeCl_3$ ((**C**): froth, (**D**): gel); (**E,F**) 20 wt % Triton X-45, 1 mmol $FeCl_3$ ((**E**): froth, (**F**): gel); (**G,H**) 50 wt % Triton X-45, 1 mmol Fe(acac)$_3$ ((**G**): froth, (**H**): gel).

With the original synthesis of extended nanosheets in the hexagonal phase of Triton X-100 in H$_2$O [47], we then decided to test if nanosheets would also be obtained in the same surfactant–H$_2$O system, but at higher surfactant concentrations (~70 wt %) where Triton X-100 forms an Lα phase, just like Triton X-45 in the syntheses described above. In the case of Triton X-100, the Lα phase is observed only below room temperature up to about 8 °C (see phase diagram by Beyer et al. [52]).

The "crumpled" nanosheets (especially those shown in Figure 4A,B) closely resembled the product obtained in the absence of a surfactant described earlier. Significantly smaller sheets, which overlap and appear to fold into each other, can be seen in the representative TEM images shown in Figure 4. Their lateral dimensions rarely exceeded a few hundred nanometers, as can be seen in Figure 4C.

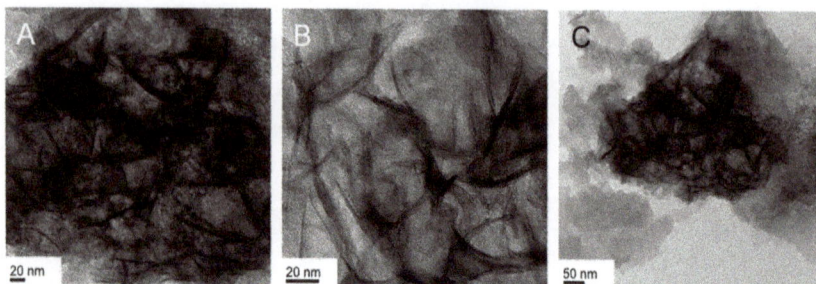

Figure 4. TEM images of the IO nanostructures obtained in the Lα phase of Triton X-100 in water (70 wt % Triton X-100): the reaction was performed at 3 °C and both gel and froth fraction contain the same product (see images in (**A–C**)).

The next surfactant system we investigated was Brij-C10. Brij-C10 in H$_2$O gives rise to a variety of LLC phases depending on the surfactant's concentration. Homogeneous (non-biphasic) LLC phases form above 35 °C, starting with the normal hexagonal columnar (Col$_{hI}$ or H$_I$) over the bicontinuous cubic (V$_I$ or Cub$_I$) to the Lα phase as the concentration of Brij-C10 is raised from about 30 to approximately 80 wt %. As described by others [54], neither the addition of the iron precursors nor the addition of the NaBH$_4$ solution leads to a significant change of the overall phase diagram. Starting again with the synthesis in the Lα phase, we first examined two particular concentrations of Brij-C10 in H$_2$O (72.5 and 55 wt %). To retain the Lα phase for the lower concentration of Brij-C10 in H$_2$O, the temperature for the second synthesis needed to be raised from 62 to 85 °C, since at this concentration the bicontinuous cubic phase below the Lα phase persisted until 78 °C.

Figures 5 and 6 show representative TEM images of both syntheses and the characteristic products obtained from both gel and froth fraction. The first key observation when analyzing these two sets of TEM images is that the products (i.e., the morphology of the obtained nanostructures) are very different despite the fact that the differences between both syntheses are only the surfactant concentration and the reaction temperature. The second important observation is that the synthesis at higher surfactant concentration (and lower temperature) produced two entirely different product morphologies, nanoparticles in the size regime of 10 nm in the froth fraction (Figure 5A,B) and the earlier (for Triton X-100 in the Lα phase, Figure 4) described 'crumpled' nanosheets (see Figure 5C,D).

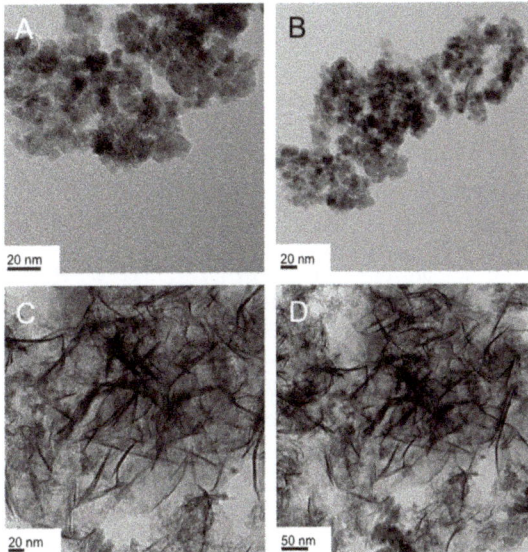

Figure 5. TEM images of the IO nanostructures obtained from the synthesis in the Lα phase formed by Brij-C10 (72.5 wt %) in water at 62 °C: (**A,B**) froth fraction and (**C,D**) gel fraction.

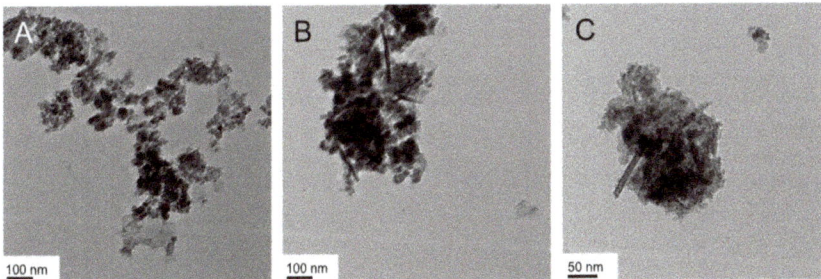

Figure 6. TEM images of the IO nanostructures obtained from the synthesis in the Lα phase formed by Brij-C10 (55 wt %) in water at 85 °C: (**A,B**) froth fraction and (**C**) gel fraction.

At higher temperatures and lower Brij-C10 concentration in the Lα phase, however, froth and gel fractions simultaneously show multiple IO nanostructure morphologies. Smaller flat sheets, even rod-like nanoparticles, and larger polyhedral nanoparticles (up to 40 nm in diameter) can be seen in both the froth and the gel fraction (Figure 6A–C). It appears that at this temperature the reaction is simply too fast to result in products with consistent shape or morphology.

We also tested the synthesis in the hexagonal columnar phase of Brij-C10 in H$_2$O and obtained extended large nanosheets as in the case of Triton X-100 (TEM images similar to those published earlier [47] and shown here in Figure 1B). As expected, the bicontinuous cubic phase was too viscous for the reaction to proceed at all.

To examine the role of the surfactant type (ionic vs. neutral), the final set of syntheses was performed in the micellar solution and hexagonal columnar LLC phase of an ionic surfactant (cetyltrimethylammonium bromide, CTAB) instead of the non-ionic oligoethyleneglycol-based Triton X or Brij series surfactants. CTAB in H$_2$O gives rise to a phase diagram similar to Brij-C10 with a normal phase sequence of isotropic (micellar, Iso)–Col$_{Hi}$–Cub$_I$–Lα as the concentration of CTAB increases.

As shown in the phase diagram reported by Raman et al., at lower concentration of CTAB the shape of the micelles in the two micellar phases changes from quasi-spherical to rod-like, giving rise to two critical micelle concentrations, CMC_1 and CMC_2 [55]. We will initially discuss the two cases of the reduction/hydrolysis reaction in the columnar phase, at first using $FeCl_3$ and then the one using $Fe(acac)_3$ as iron precursor, each at 1 mmol.

We can see right away that the froth (Figure 7A–C) and gel fractions (Figure 7D–F) of the hydrolysis/reduction of $FeCl_3$ as iron precursor in the Col_{hI} phase produce entirely different IO nanoscale morphologies. The froth fraction contains what appear to be nanodiscs, some more round others square with rounded edges. The assignment of a disc shape rests on the numerous overlapping and highly crystalline structures for which clear lattice fringes can be seen as, for example, in Figure 7B. The average lateral dimensions of these structures are about 20 nm. The gel fraction, however, contained clearly discernable nanosheets with some smaller nanostructures (somewhat similar to those found in the froth fraction in both size and shape, see Figure 7E). These nanosheets resemble the 'crumpled' nanosheets discussed earlier and some are oriented to allow an estimation of the height of these nanosheets to be ~5 or ~10 nm. The lateral dimensions are smaller than those initially obtained for the synthesis in the Col_{hI} phase formed by Triton X-100 in H_2O [47].

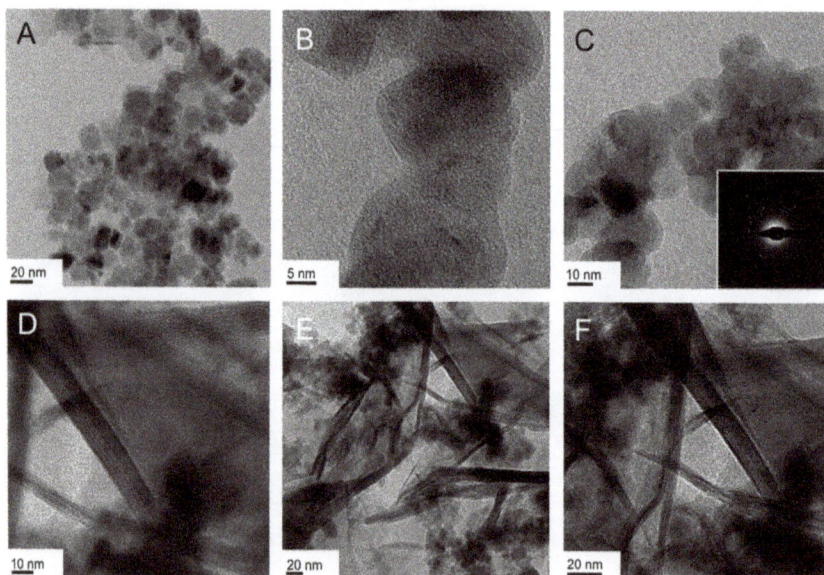

Figure 7. TEM images of the IO nanostructures obtained from the reduction/hydrolysis synthesis of $FeCl_3$ in the Col_{hI} phase formed by CTAB (25 wt %) in water at 25 °C: (**A–C**) froth fraction and (**D–F**) gel fraction. The inset in (**C**) shows the selected area electron diffraction (SAED) pattern, which confirms the crystallinity of these quasi-circular nanodiscs.

Suprisingly, the same reaction at the same conditions now using $Fe(acac)_3$ as iron precursor did not have the same outcome with respect to significant differences in shape in the froth and gel fractions.

As can be seen in the TEM images collected in Figure 8, the froth (Figure 8A,B) contains rough nanoparticle-like structures whose specific shape is not easy to determine from the TEM images. The size of these particles ranges from about 10 to 20 nm. The gel fraction now contained what we previously suggested to be nanodiscs—highly crystalline (lattice fringes are visible at closer inspection), strongly overlapping quasi-spherical shapes with about 10 to 20 nm lateral dimensions (Figure 8C,D). In essence, merely changing the type of iron(III) precuror (or just its counterion) alters two reaction outcomes: (i) what type of morphology is found in froth or gel fraction and (ii) the complete disappearance of nanosheets as a reaction product. Notably, the same also occurred for the related syntheses in the Lα phase of Triton X-45 shown in Figure 3G,H. In fact, the products of that reaction earlier described are very similar to the products of the froth fraction obtained here for CTAB (Figure 8A,B).

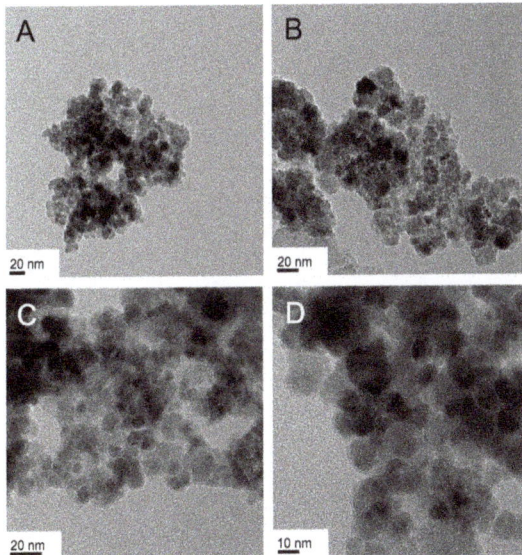

Figure 8. TEM images of the IO nanostructures obtained from the reduction/hydrolysis reaction of Fe(acac)$_3$ in the Col$_{hI}$ phase formed by CTAB (25 wt %) in water at 25 °C: (**A,B**) froth fraction and (**C,D**) gel fraction (lattice fringes are visible for some of the nanostructures in image (**D**)).

Since the Lα phase formed by CTAB in H$_2$O only forms at surfactant concentrations above 70 or 80 wt % depending on the temperature (Note: also the Krafft temperature—the temperature at which micelles form—increases markedly for higher concentrations of CTAB in H$_2$O), the viscosity of the Lα phase was too high and the solubility of the iron(III) precursors too limited to pursure the reaction. What we did test is the effect of the micelle shape on the reaction outcome by running the reaction in two concentrations of CTAB that form spherical or rod-like micelles (i.e., above CMC_1 and above CMC_2), respectively.

At the lower concentration of CTAB in H$_2$O (3.6 wt %, i.e., in the aqueous solution featuring spherical micelles) the froth fraction contained, as in several syntheses before, 'crumpled' sheets, consistently thin with an average height of about 5 nm (Figure 9A–C). The gel fraction (Figure 9D–F) contained two distinct products with respect to shape, which in some cases could be separated by meticulous washing, magnetic separation and/or centrifugation. As shown in the TEM images, the two shapes are either nanoparticles with a quite significant size distribution (ranging from 10 to 50 nm in diameter) as well as extended nanosheets (Figure 9E) with lateral dimensions of several hundred nanometers.

The same experiment performed in the isotropic phase consisting of rod-like micelles (i.e., at 20 wt % CTAB in H$_2$O, above CMC_2) led to the formation of highly "crumpled" nanosheets in the froth fraction (Figure 10A,B) and flower-like nanoparticles in the size regime of 10 to 20 nm in the gel fraction (Figure 10C,D).

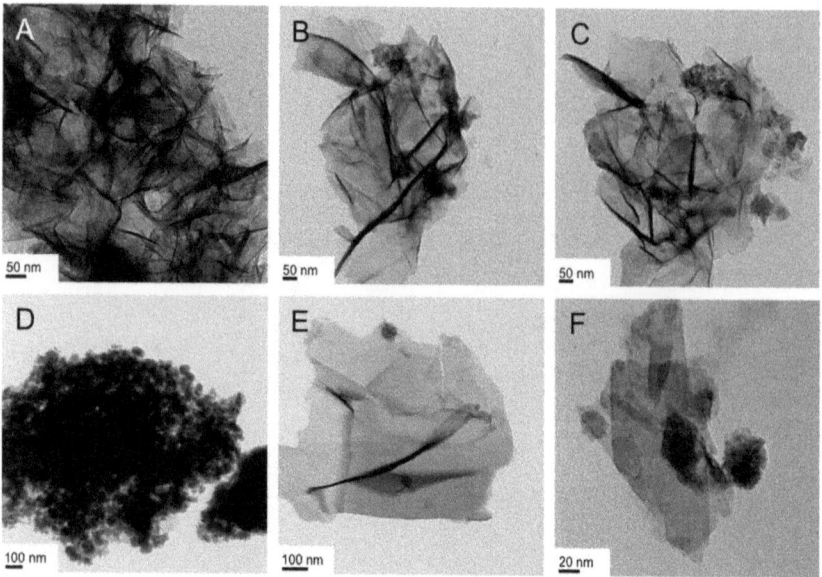

Figure 9. TEM images of the IO nanostructures obtained from the reduction/hydrolysis reaction of FeCl$_3$ in the isotropic micellar liquid (spherical micelles) formed by CTAB (3.6 wt %) in water at 50 °C: (**A**–**C**) froth fraction and (**D**–**F**) gel fraction.

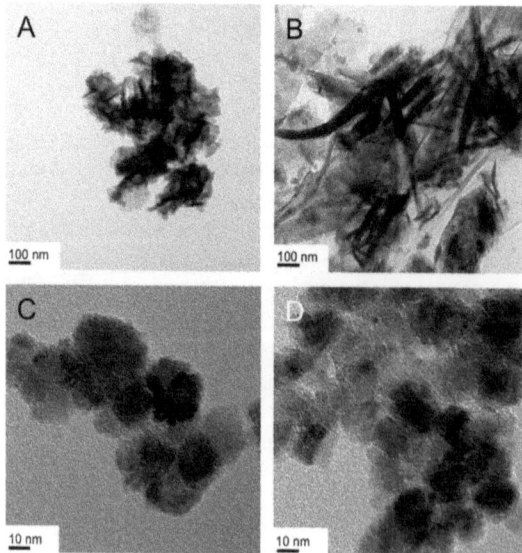

Figure 10. TEM images of the IO nanostructures obtained from the reduction/hydrolysis reaction of FeCl$_3$ in the isotropic micellar liquid (rod-like micelles) formed by CTAB (20 wt %) in water at 60 °C: (**A**,**B**) froth fraction and (**C**,**D**) gel fraction.

It appears that altering the shape of the micelles in solution (by altering the concentration of the surfactant) affects the nanoscale morphology of the IO nanomaterial shape formed during the reduction/hydrolysis reaction.

Finally, exchanging the iron(III) precursor from $FeCl_3$ to $Fe(acac)_3$ led again to the rough-edged quasi-spherical nanoparticles (Figure 11) obtained earlier from the reduction/hydrolysis reaction of $Fe(acac)_3$ in the Col_{hI} phase formed by CTAB in H_2O (see Figure 8A,B). Considering our original report of this reduction/hydrolysis reaction to obtain pure magnetite (Fe_3O_4) nanoparticles using $Fe(acac)_3$ as sole iron precursor [51], it appears that the acac ligand, perhaps because of its effect on the iron(III) precursor's solubility or reactivity, favors the formation of quasi-spherical nanoparticles over flat nanosheet or nanodisc-like morphologies as in the case of iron(III) precursors with chloride counter-ions. The reaction is in both cases incredibly fast (a few minutes), but it is conceivable that acac as bidentate ligand binds to initially-formed nanoparticles, preventing any coalescence to larger, flat sheet- or disc-like nanostructures.

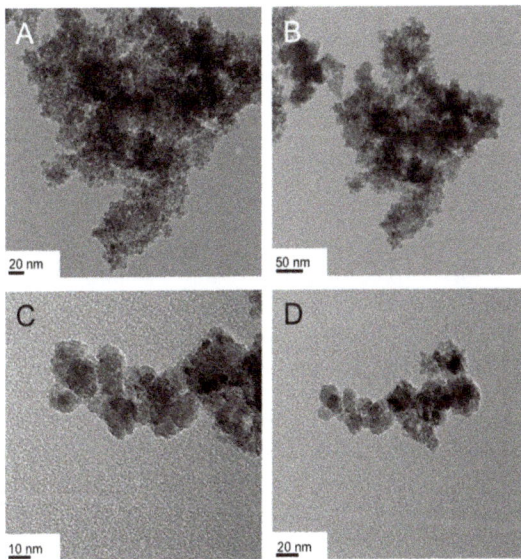

Figure 11. TEM images of the IO nanostructures obtained from the reduction/hydrolysis reaction of $Fe(acac)_3$ in the isotropic micellar liquid (rod-like micelles) formed by CTAB (20 wt %) in water at 60 °C: (**A,B**) froth fraction and (**C,D**) gel fraction.

3. Discussion

Table 1 summarizes the size- and shape-dependencies of the IO nanomaterials obtained from the syntheses in the micellar solutions and bulk lyotropic LC phases serving as solvents, depending also on the fraction (froth, gel) from which they were isolated. Particularly astonishing is the fact that even a very subtle change in some of the surfactant systems results in tremendously different IO nanomaterial shapes. For example, doubling the concentration of the iron precursor in the Triton X-45 Lα phase results in nanosheets with larger lateral dimensions, and changing the type of iron precursor brings about significant shape changes (nanosheets vs. NPs in the gel fraction). Changing the type of Triton X surfactant and temperature, but performing the synthesis in both cases in the Lα phase, alters the type of nanosheet (flat and extended vs. crumpled nanosheets). Moving on to another oligoethylene glycol-based surfactant, Brij-C10, and to keep the Lα phase at higher reaction temperatures introduces a more significant amount of polyhedral NPs. This trend appears to be somewhat consistent among the various surfactant systems, in that higher temperatures seem to favor the formation of smaller nanostructures (NPs or nanodiscs) rather than larger extended or crumpled nanosheets.

Table 1. Summary of nanostructures formed in the various micellar solutions and LLC phases.

Surfactant	Iron Precursor/Conc., Phase/wt % Surfactant, Temperature [1]	Main Morphology [2] Gel [Froth]	Dimensions [3] $w \times d \times h$ (nm)
Triton X-45	$FeCl_3$/1 mmol, Lα/50, 40 °C	NS [NS + NPs]	100 × 100 × 5 [20]
	$FeCl_3$/2 mmol, Lα/50, 40 °C	NS + NPs [NS + NPs]	300 × 300 × 10 [20]
	$FeCl_3$/1 mmol, Lα/20, 40 °C	NP [NS]	300 × 300 × 10 [20]
	$Fe(acac)_3$/1 mmol, Lα/50, 40 °C	NP [NP]	15–20 [15–20]
Triton X-100	$FeCl_3$/1 mmol, Col_hI/48, 25 °C [47]	NS [NS]	500+ × 500+ × 5, 10
	$FeCl_3$/1 mmol, Lα/70, 3 °C	NS_cr [NS_cr]	100+ × 100+ × 5, 10
Brij-C10	$FeCl_3$/1 mmol, Lα/72.5, 62 °C	NS_cr [NP]	10 [100+ × 100+ × 5, 10]
	$FeCl_3$/1 mmol, Lα/55, 85 °C	NP + few NS_cr [NP]	40 [40]
CTAB	$FeCl_3$/1 mmol, Col_hI/25, 25 °C	NS + NP [ND]	100 × 100 × 10 [20] [4]
	$Fe(acac)_3$/1 mmol, Col_hI/25, 25 °C	ND [NP]	20 [4] [20]
	$FeCl_3$/1 mmol, Iso_sp-mic/3.6, 50 °C	NS + NP [NS_cr]	500+ × 500+ × 10, 20
	$FeCl_3$/1 mmol, Iso_rl-mic/20, 60 °C	ND [NS_cr]	20–30 [4] [100+ × 100+ × 20]
	$Fe(acac)_3$/1 mmol, Iso_rl-mic/20, 60 °C	NP [NP]	10–20 [10–20]

[1] Parameter adjusted in bold font; [2] NS = nanosheet, NS_cr = "crumpled" NS, NP = nanoparticle, ND = nanodisc, Iso_sp-mic = isotropic liquid containing spherical micelles, and Iso_rl-mic = isotropic liquid containing rod-like micelles; [3] Where applicable for polyhedral shapes; otherwise the average diameter of the polyhedral NPs is given. [4] The diameter of the ND is given.

The Triton X-100 vs. CTAB syntheses in the Col_hI bulk LC phase exemplify the significance of the nature of the surfactant (i.e., neutral vs. ionic): the ionic CTAB typically favors the formation of nanodiscs, which were not observed in the synthesis using Triton X-100. The shape of the micelles in the CTAB isotropic system (above CMC_1 and CMC_2) also shows some effect with no formation of polyhedral nanoparticles in the case of rod-like micelles, unless $Fe(acac)_3$ is used as the iron precursor. As a more general observation, for the discussed reduction-hydrolysis reaction, $Fe(acac)_3$ in almost all cases favors the formation of polyhedral, quasi-spherical NPs instead of nanosheets or nanodiscs favored when $FeCl_3$ is the iron precursor.

This is consistent with our earlier observation of the formation of NPs from $Fe(acac)_3$ in our reduction-hydrolysis reaction [51], which might be due to the fact that the reduced form of acac, i.e., pentane-2,4-diol [56], could act as a stabilizing bidendate ligand preventing the coalescence of initially formed nanocrystallites that form the basis of the larger nanosheets formed in the case of $FeCl_3$ as iron precursor. As for the mechanism of the formation of nanosheets and nanodiscs, the dimensions of the lamellar sheets in the Lα phases and the rod-like micelles in the Col_hI phases appear to play a signficant role.

As described before, the height of the flat nanostructures always appears to be a value close to or multiple-times the layer spacing (Lα) or lattice paramater (Col_hI) of the lyotropic LC phase used as a solvent [47], as determined by TEM tomography (for an example see Figure 12). We therefore argued that confinement in the growth of IO nanostructures within the lamellae or columnar aggregates of the LLC structure, as well as confinement on the gas bubbles forming the froth, play important roles in the formation of the various IO nanostructures [47]. However, as exemplified by the synthesis that uses the Lα phase of Brij-C10 (Figure 5), significant differences in nanostructure morphology in froth and gel fractions cannot solely be explained by confinement on the froth's gas bubbles. In most other syntheses involving oligoethylene glycol-based surfactant systems (i.e., Triton X-45 and X-100), nanosheets were predominantly formed in the froth and not in the gel fraction. In the case of Brij-C10 (at 72.5 wt % in H_2O and 62 °C, Figure 5A–D), "crumpled" nanosheets formed exclusively in the gel fraction, perhaps due to the higher viscosity and the resulting lower rate of IO nanostructure formation. Hence, a conclusive general mechanism of IO nanostructure formation in these LLC solvents is rather difficult to deduce considering additional effects such as coordination of atoms or groups of atoms of the various surfactant molecules to the IO nanostructure surfaces (e.g., oxygen atoms in the oligoethylene glycol segments) and differences in viscosity that affect the rate of IO nanostructure formation. The viscosity of a given LLC solvent system is governed by various parameters such as temperature, surfactant and precursor concentration, as well as the type of micellar or LLC phase. In this sense, the LLC phases are not simply passive templates but rather complex anisometric as well as coordinating solvents that affect both shape and dimensions of the formed IO nanomaterials in very specific ways depending on all synthesis parameters.

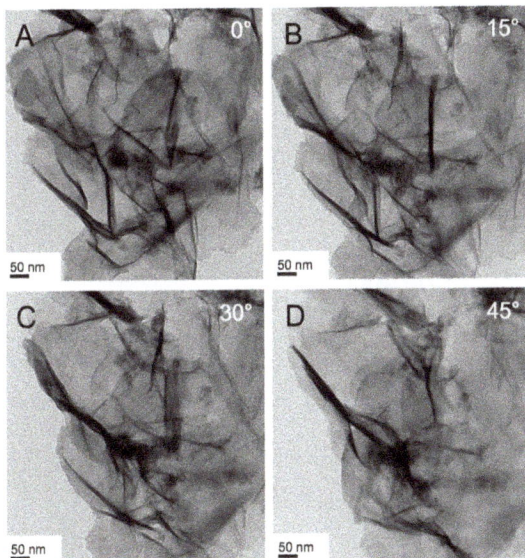

Figure 12. TEM tomography images (various degrees of tilt) of crumpled IO nanosheets obtained via the reduction/hydrolysis reaction of $FeCl_3$ in the isotropic micellar liquid (spherical micelles) formed by CTAB (3.6 wt %) in water at 50 °C (froth fraction).

Considering earlier data on the differential cell internalization of polyhedral IO nanobricks vs. quasi-spherical IO NPs [42], future investigations of cell uptake by endothelial as well as epithelial cells will focus particularly on the smaller nanodiscs obtained from several of the CTAB–H_2O LLC solvent systems. These comparative studies will allow us to expand our understanding of the role of NP shape in regulating preferential uptake of NPs over a wider range of IO nanomaterial shapes, including nanospheres, -bricks, and -discs.

4. Materials and Methods

Iron(III) chloride hexahydrate ($FeCl_3 \cdot 6H_2O$, 98%), iron(III) acetylacetonate ($Fe(acac)_3$, 97%,), sodium borohydride ($NaBH_4$, 98%), cetyltrimethylammonium bromide (CTAB, $\geq$99%), Triton™ X-100 (laboratory grade), and Triton™ X-45, Brij® C10 were purchased from Sigma-Aldrich (St. Louis, MO, US)and used as such for synthesis. Deionized (DI) water (R = 18 MΩ) was used in all reaction and purification steps. Transmission electron microscopy (TEM) was carried out using a FEI Tecnai TF20 TEM instrument (Hillsboro, OR, USA) at an accelerating voltage of 200 kV. Particle samples were dispersed in water or methanol and a droplet was placed onto a carbon coated copper grid (400 mesh) and finally air-dried prior to analysis. Polarized Optical Microscopy (POM) studies were carried out using an Olympus BX-53 microscope equipped with a Linkam LTS420E (Waterfield, UK) heating/cooling stage. For each reaction, small samples were taken before and after the addition of the reducing agent (NaBH$_4$), and the LLC phase formed and retained was confirmed by careful texture observations between crossed polarizers (for an example, see Figure 2D,E).

The surfactant weighed into the reaction flask was degassed under high vacuum for two hours. Separately, FeCl$_3$·6H$_2$O (1 or 2 mmol) or alternatively Fe(acac)$_3$ in the appropriate amount of DI water (to obtain the LLC phase or micellar solution of interest to a total of about 50 mL) was purged with nitrogen for one hour. Thereafter, the FeCl$_3$ or Fe(acac)$_3$ solution was added to the surfactant in the reaction flask under a steady flow of nitrogen, which was maintained until the completion of the reaction. The contents were mixed using a mechanical stirrer at 100 RPM, until the mixture turned to a taffy-like consistency. At this stage, the contents were slowly warmed to 40 or 50 °C (until all solids went into solution to form the micellar solution or bulk LLC phase) and cooled back to ambient temperature to form a yellow-colored bulk LLC phase (checked by POM). NaBH$_4$ (10 or 20 mmol) in 5 or 10 mL deoxygenated DI water was added rapidly to the reaction vessel with occasional mechanical stirring. The amount of water to prepare the NaBH$_4$ solution was taken into account when calculating the surfactant concentrations to obtain specific micellar solutions or LLC phases. The yellow gel turned almost immediately to a black color along with formation of excess froth that filled the reaction vessel. Although the reaction is complete after approximately two minutes, after an additional 30 min from adding NaBH$_4$, the black-colored froth was collected by adding 200 mL DI water in portions, leaving the gel part behind. The crude product from the froth was collected using a rare earth magnet and washed several times with DI water and followed by ethanol. The final product after washing was dried at room temperature in air to yield a black powder. The product isolated from the gel portion after washing with copious amounts of water followed by ethanol was a dark blackish powder. This synthesis procedure was followed for all surfactants and surfactant concentrations—amounts, ratios, and concentrations are given in the captions of the TEM images.

5. Conclusions

We have shown that surfactant-water systems, both as bulk lyotropic LC or micellar solutions, are noteworthy solvents for the shape- and size-controlled synthesis of IO nanostructures. Using the reduction-hydrolysis reaction of a single iron(III) precursor discovered in our laboratory we were able to synthesize a range of magnetic IO nanostructures with disc, sheet, crumpled sheet, as well as polyhedral shapes. Based on the datasets gathered from TEM image analysis we were also able to establish how parameters such as type and concentration of the iron(III) precursor, LLC phase, type of surfactant, and temperature affect the synthesis outcome, especially with respect to the shape of the IO nanostructures. We believe that these results bode well for continued studies on template-based syntheses of metal and metal oxide nanomaterials using LLC phases. As outlined in the introduction, especially for IO nanomaterials, shape is of critical importance for differential cell uptake, hyperthermia, MRI contrast enhancement, and drug delivery. For metal oxides in general, the large range of possible, already considered and tested applications aside from medical uses will certainly benefit from synthetic tools that allow for control over nanomaterial size and shape.

Supplementary Materials: The following are available online at http://www.mdpi.com/2079-4991/7/8/211/s1.

Acknowledgments: This work was supported by the Ohio Third Frontier (OTF) program for Ohio Research Scholars "Research Cluster on Surfaces in Advanced Materials" (support for T.H.), which also supports the cryo-TEM facility at the Liquid Crystal Institute (KSU), where current TEM data were acquired.

Author Contributions: Torsten Hegmann and Donald W. Miller conceived and Torsten Hegmann, Seyyed Muhammad Salili and Matthew Worden designed the experiments; Seyyed Muhammad Salili and Matthew Worden performed the experiments; Torsten Hegmann, Seyyed Muhammad Salili, Matthew Worden and Ahlam Nemati analyzed the data; Torsten Hegmann wrote the paper.

Conflicts of Interest: The authors declare no conflict of interest.

References

1. Kurzhals, S.; Gal, N.; Zirbs, R.; Reimhult, E. Controlled aggregation and cell uptake of thermoresponsive polyoxazoline-grafted superparamagnetic iron oxide nanoparticles. *Nanoscale* **2017**, *9*, 2793–2805. [CrossRef] [PubMed]

2. Lassenberger, A.; Scheberl, A.; Stadlbauer, A.; Stiglbauer, A.; Helbich, T.; Reimhult, E. Individually stabilized, superparamagnetic nanoparticles with controlled shell and size leading to exceptional stealth properties and high relaxivities. *ACS Appl. Mater. Interfaces* **2017**, *9*, 3343–3353. [CrossRef] [PubMed]

3. Gal, N.; Lassenberger, A.; Herrero-Nogareda, L.; Scheberl, A.; Charwat, V.; Kasper, C.; Reimhult, E. Interaction of size-tailored pegylated iron oxide nanoparticles with lipid membranes and cells. *ACS Biomater. Sci. Eng.* **2017**, *3*, 249–259. [CrossRef]

4. Ghosh, D.; Lee, Y.; Thomas, S.; Kohli, A.G.; Yun, D.S.; Belcher, A.M.; Kelly, K.A. M13-templated magnetic nanoparticles for targeted in vivo imaging of prostate cancer. *Nat. Nanotechnol.* **2012**, *7*, 677–682. [CrossRef] [PubMed]

5. Bertoli, F.; Davies, G.L.; Monopoli, M.P.; Moloney, M.; Gun'ko, Y.K.; Salvati, A.; Dawson, K.A. Magnetic nanoparticles to recover cellular organelles and study the time resolved nanoparticle-cell interactome throughout uptake. *Small* **2014**, *10*, 3307–3315. [CrossRef] [PubMed]

6. Basuki, J.S.; Esser, L.; Duong, H.T.T.; Zhang, Q.; Wilson, P.; Whittaker, M.R.; Haddleton, D.M.; Boyer, C.; Davis, T.P. Magnetic nanoparticles with diblock glycopolymer shells give lectin concentration-dependent mri signals and selective cell uptake. *Chem. Sci.* **2014**, *5*, 715–726. [CrossRef]

7. Grafe, C.; Slabu, I.; Wiekhorst, F.; Bergemann, C.; von Eggeling, F.; Hochhaus, A.; Trahms, L.; Clement, J.H. Magnetic particle spectroscopy allows precise quantification of nanoparticles after passage through human brain microvascular endothelial cells. *Phys. Med. Biol.* **2016**, *61*, 3986–4000. [CrossRef] [PubMed]

8. Davila-Ibanez, A.B.; Salgueirino, V.; Martinez-Zorzano, V.; Marino-Fernandez, R.; Garcia-Lorenzo, A.; Maceira-Campos, M.; Munoz-Ubeda, M.; Junquera, E.; Aicart, E.; Rivas, J.; et al. Magnetic silica nanoparticle cellular uptake and cytotoxicity regulated by electrostatic polyelectrolytes-DNA loading at their surface. *ACS Nano* **2012**, *6*, 747–759. [CrossRef] [PubMed]

9. Hofmann, D.; Tenzer, S.; Bannwarth, M.B.; Messerschmidt, C.; Glaser, S.F.; Schild, H.; Landfester, K.; Mailander, V. Mass spectrometry and imaging analysis of nanoparticle-containing vesicles provide a mechanistic insight into cellular trafficking. *ACS Nano* **2014**, *8*, 10077–10088. [CrossRef] [PubMed]

10. Levy, M.; Gazeau, F.; Bacri, J.C.; Wilhelm, C.; Devaud, M. Modeling magnetic nanoparticle dipole-dipole interactions inside living cells. *Phys. Rev. B* **2011**, *84*, 075480. [CrossRef]

11. Lunov, O.; Zablotskii, V.; Syrovets, T.; Rocker, C.; Tron, K.; Nienhaus, G.U.; Simmet, T. Modeling receptor-mediated endocytosis of polymer-functionalized iron oxide nanoparticles by human macrophages. *Biomaterials* **2011**, *32*, 547–555. [CrossRef] [PubMed]

12. Fernandes, A.R.; Chari, D.M. A multicellular, neuro-mimetic model to study nanoparticle uptake in cells of the central nervous system. *Integr. Biol.* **2014**, *6*, 855–861. [CrossRef] [PubMed]

13. Bothun, G.D.; Lelis, A.; Chen, Y.J.; Scully, K.; Anderson, L.E.; Stoner, M.A. Multicomponent folate-targeted magnetoliposomes: Design, characterization, and cellular uptake. *Nanomed. Nanotechnol. Biol. Med.* **2011**, *7*, 797–805. [CrossRef] [PubMed]

14. Liong, M.; Lu, J.; Kovochich, M.; Xia, T.; Ruehm, S.G.; Nel, A.E.; Tamanoi, F.; Zink, J.I. Multifunctional inorganic nanoparticles for imaging, targeting, and drug delivery. *ACS Nano* **2008**, *2*, 889–896. [CrossRef] [PubMed]

15. Lartigue, L.; Wilhelm, C.; Servais, J.; Factor, C.; Dencausse, A.; Bacri, J.C.; Luciani, N.; Gazeau, F. Nanomagnetic sensing of blood plasma protein interactions with iron oxide nanoparticles: Impact on macrophage uptake. *ACS Nano* **2012**, *6*, 2665–2678. [CrossRef] [PubMed]

16. Zhu, K.N.; Deng, Z.Y.; Liu, G.H.; Hu, J.M.; Liu, S.Y. Photoregulated cross-linking of superparamagnetic iron oxide nanoparticle (spion) loaded hybrid nanovectors with synergistic drug release and magnetic resonance (MR) imaging enhancement. *Macromolecules* **2017**, *50*, 1113–1125. [CrossRef]

17. Sherlock, S.P.; Tabakman, S.M.; Xie, L.M.; Dai, H.J. Photothermally enhanced drug delivery by ultrasmall multifunctional feco/graphitic shell nanocrystals. *ACS Nano* **2011**, *5*, 1505–1512. [CrossRef] [PubMed]

18. Zhang, S.L.; Gao, H.J.; Bao, G. Physical principles of nanoparticle cellular endocytosis. *ACS Nano* **2015**, *9*, 8655–8671. [CrossRef] [PubMed]

19. Jeon, S.; Hurley, K.R.; Bischof, J.C.; Haynes, C.L.; Hogan, C.J. Quantifying intra- and extracellular aggregation of iron oxide nanoparticles and its influence on specific absorption rate. *Nanoscale* **2016**, *8*, 16053–16064. [CrossRef] [PubMed]

20. Jing, Y.; Mal, N.; Williams, P.S.; Mayorga, M.; Penn, M.S.; Chalmers, J.J.; Zborowski, M. Quantitative intracellular magnetic nanoparticle uptake measured by live cell magnetophoresis. *FASEB J.* **2008**, *22*, 4239–4247. [CrossRef] [PubMed]

21. Gupta, A.K.; Naregalkar, R.R.; Vaidya, V.D.; Gupta, M. Recent advances on surface engineering of magnetic iron oxide nanoparticles and their biomedical applications. *Nanomedicine* **2007**, *2*, 23–39. [CrossRef] [PubMed]

22. Xu, Y.L.; Sherwood, J.A.; Lackey, K.H.; Qin, Y.; Bao, Y.P. The responses of immune cells to iron oxide nanoparticles. *J. Appl. Toxicol.* **2016**, *36*, 543–553. [CrossRef] [PubMed]

23. Sherwood, J.; Lovas, K.; Rich, M.; Yin, Q.; Lackey, K.; Bolding, M.S.; Bao, Y. Shape-dependent cellular behaviors and relaxivity of iron oxide-based T-1 mri contrast agents. *Nanoscale* **2016**, *8*, 17506–17515. [CrossRef] [PubMed]

24. Hirsch, V.; Kinnear, C.; Moniatte, M.; Rothen-Rutishauser, B.; Clift, M.J.D.; Fink, A. Surface charge of polymer coated spions influences the serum protein adsorption, colloidal stability and subsequent cell interaction in vitro. *Nanoscale* **2013**, *5*, 3723–3732. [CrossRef] [PubMed]

25. Mo, J.B.; He, L.Z.; Ma, B.; Chen, T.F. Tailoring particle size of mesoporous silica nanosystem to antagonize glioblastoma and overcome blood-brain barrier. *ACS Appl. Mater. Interfaces* **2016**, *8*, 6811–6825. [CrossRef] [PubMed]

26. Kievit, F.M.; Stephen, Z.R.; Veiseh, O.; Arami, H.; Wang, T.Z.; Lai, V.P.; Park, J.O.; Ellenbogen, R.G.; Disis, M.L.; Zhang, M.Q. Targeting of primary breast cancers and metastases in a transgenic mouse model using rationally designed multifunctional spions. *ACS Nano* **2012**, *6*, 2591–2601. [CrossRef] [PubMed]

27. Xu, C.J.; Miranda-Nieves, D.; Ankrum, J.A.; Matthiesen, M.E.; Phillips, J.A.; Roes, I.; Wojtkiewicz, G.R.; Juneja, V.; Kultima, J.R.; Zhao, W.A.; et al. Tracking mesenchymal stem cells with iron oxide nanoparticle loaded poly(lactide-*co*-glycolide) microparticles. *Nano Lett.* **2012**, *12*, 4131–4139. [CrossRef] [PubMed]

28. Massart, R. Preparation of aqueous magnetic liquids in alkaline and acidic media. *IEEE Trans. Magn.* **1981**, *17*, 1247–1248. [CrossRef]

29. Sun, S.H.; Zeng, H. Size-controlled synthesis of magnetite nanoparticies. *J. Am. Chem. Soc.* **2002**, *124*, 8204–8205. [CrossRef] [PubMed]

30. Mendoza-Garcia, A.; Sun, S.H. Recent advances in the high-temperature chemical synthesis of magnetic nanoparticles. *Adv. Funct. Mater.* **2016**, *26*, 3809–3817. [CrossRef]

31. Song, M.M.; Song, W.J.; Bi, H.; Wang, J.; Wu, W.L.; Sun, J.; Yu, M. Cytotoxicity and cellular uptake of iron nanowires. *Biomaterials* **2010**, *31*, 1509–1517. [CrossRef] [PubMed]

32. Gil, S.; Correia, C.R.; Mano, J.F. Magnetically labeled cells with surface-modified Fe$_3$O$_4$ spherical and rod-shaped magnetic nanoparticles for tissue engineering applications. *Adv. Healthc. Mater.* **2015**, *4*, 883–891. [CrossRef] [PubMed]

33. Yu, P.; Xia, X.M.; Wu, M.; Cui, C.; Zhang, Y.; Liu, L.; Wu, B.; Wang, C.X.; Zhang, L.J.; Zhou, X.; et al. Folic acid-conjugated iron oxide porous nanorods loaded with doxorubicin for targeted drug delivery. *Coll. Surf. B* **2014**, *120*, 142–151. [CrossRef] [PubMed]

34. Yue, Z.G.; Wei, W.; You, Z.X.; Yang, Q.Z.; Yue, H.; Su, Z.G.; Ma, G.H. Iron oxide nanotubes for magnetically guided delivery and pH-activated release of insoluble anticancer drugs. *Adv. Funct. Mater.* **2011**, *21*, 3446–3453. [CrossRef]

35. Safi, M.; Yan, M.H.; Guedeau-Boudeville, M.A.; Conjeaud, H.; Garnier-Thibaud, V.; Boggetto, N.; Baeza-Squiban, A.; Niedergang, F.; Averbeck, D.; Berret, J.F. Interactions between magnetic nanowires and living cells: Uptake, toxicity, and degradation. *ACS Nano* **2011**, *5*, 5354–5364. [CrossRef] [PubMed]

36. Wu, P.C.; Wang, W.S.; Huang, Y.T.; Sheu, H.S.; Lo, Y.W.; Tsai, T.L.; Shieh, D.B.; Yeh, C.S. Porous iron oxide based nanorods developed as delivery nanocapsules. *Chem. Eur. J.* **2007**, *13*, 3878–3885. [CrossRef] [PubMed]

37. Key, J.; Dhawan, D.; Cooper, C.L.; Knapp, D.W.; Kim, K.; Kwon, I.C.; Choi, K.; Park, K.; Decuzzi, P.; Leary, J.F. Multicomponent, peptide-targeted glycol chitosan nanoparticles containing ferrimagnetic iron oxide nanocubes for bladder cancer multimodal imaging. *Int. J. Nanomed.* **2016**, *11*, 4141–4155. [CrossRef] [PubMed]

38. Wortmann, L.; Ilyas, S.; Niznansky, D.; Valldor, M.; Arroub, K.; Berger, N.; Rahme, K.; Holmes, J.; Mathur, S. Bioconjugated iron oxide nanocubes: Synthesis, functionalization, and vectorization. *ACS Appl. Mater. Interfaces* **2014**, *6*, 16631–16642. [CrossRef] [PubMed]

39. Xiong, F.; Chen, Y.J.; Chen, J.X.; Yang, B.Y.; Zhang, Y.; Gao, H.L.; Hua, Z.C.; Gu, N. Rubik-like magnetic nanoassemblies as an efficient drug multifunctional carrier for cancer theranostics. *J. Control. Release* **2013**, *172*, 993–1001. [CrossRef] [PubMed]

40. Yuan, H.Y.; Timmerman, M.; van de Putte, M.; Rodriguez, P.G.; Veldhuis, S.; ten Elshof, J.E. Self-assembly of metal oxide nanosheets at liquid-air interfaces in colloidal solutions. *J. Phys. Chem. C* **2016**, *120*, 25411–25417. [CrossRef]

41. Gao, Z.Y.; Ma, T.C.; Zhao, E.Y.; Docter, D.; Yang, W.S.; Stauber, R.H.; Gao, M.Y. Small is smarter: Nano mri contrast agents-advantages and recent achievements. *Small* **2016**, *12*, 556–576. [CrossRef] [PubMed]

42. Sun, Z.Z.; Worden, M.; Wroczynskyj, Y.; Manna, P.K.; Thliveris, J.A.; van Lierop, J.; Hegmann, T.; Miller, D.W. Differential internalization of brick shaped iron oxide nanoparticles by endothelial cells. *J. Mater. Chem. B* **2016**, *4*, 5913–5920. [CrossRef]

43. Vacha, R.; Martinez-Veracoechea, F.J.; Frenkel, D. Receptor-mediated endocytosis of nanoparticles of various shapes. *Nano Lett.* **2011**, *11*, 5391–5395. [CrossRef] [PubMed]

44. Nangia, S.; Sureshkumar, R. Effects of nanoparticle charge and shape anisotropy on translocation through cell membranes. *Langmuir* **2012**, *28*, 17666–17671. [CrossRef] [PubMed]

45. Zhang, Y.; Tekobo, S.; Tu, Y.; Zhou, Q.F.; Jin, X.L.; Dergunov, S.A.; Pinkhassik, E.; Yan, B. Permission to enter cell by shape: Nanodisk vs. nanosphere. *ACS Appl. Mater. Interfaces* **2012**, *4*, 4099–4105. [CrossRef] [PubMed]

46. Worden, M.; Bruckman, M.A.; Kim, M.H.; Steinmetz, N.F.; Kikkawa, J.M.; LaSpina, C.; Hegmann, T. Aqueous synthesis of polyhedral "brick-like" iron oxide nanoparticles for hyperthermia and t-2 mri contrast enhancement. *J. Mater. Chem. B* **2015**, *3*, 6877–6884. [CrossRef] [PubMed]

47. Yathindranath, V.; Ganesh, V.; Worden, M.; Inokuchi, M.; Hegmann, T. Highly crystalline iron/iron oxide nanosheets via lyotropic liquid crystal templating. *RSC Adv.* **2013**, *3*, 9210–9213. [CrossRef]

48. Worden, M.; Bergquist, L.; Hegmann, T. A quick and easy synthesis of fluorescent iron oxide nanoparticles featuring a luminescent carbonaceous coating via in situ pyrolysis of organosilane ligands. *RSC Adv.* **2015**, *5*, 100384–100389. [CrossRef]

49. Yathindranath, V.; Worden, M.; Sun, Z.Z.; Miller, D.W.; Hegmann, T. A general synthesis of metal (Mn, Fe, Co, Ni, Cu, Zn) oxide and silica nanoparticles based on a low temperature reduction/hydrolysis pathway. *RSC Adv.* **2013**, *3*, 23722–23729. [CrossRef]

50. Yathindranath, V.; Sun, Z.Z.; Worden, M.; Donald, L.J.; Thliveris, J.A.; Miller, D.W.; Hegmann, T. One-pot synthesis of iron oxide nanoparticles with functional silane shells: A versatile general precursor for conjugations and biomedical applications. *Langmuir* **2013**, *29*, 10850–10858. [CrossRef] [PubMed]

51. Yathindranath, V.; Rebbouh, L.; Moore, D.F.; Miller, D.W.; van Lierop, J.; Hegmann, T. A versatile method for the reductive, one-pot synthesis of bare, hydrophilic and hydrophobic magnetite nanoparticles. *Adv. Funct. Mater.* **2011**, *21*, 1457–1464. [CrossRef]

52. Beyer, K. Phase structures, water binding, and molecular-dynamics in liquid-crystalline and frozen states of the system Triton X-100-D$_2$O—A deuteron and carbon NMR-study. *J. Coll. Interface Sci.* **1982**, *86*, 73–89. [CrossRef]

53. Fritscher, C.; Husing, N.; Bernstorff, S.; Brandhuber, D.; Koch, T.; Seidler, S.; Lichtenegger, H.C. In situ saxs study on cationic and non-ionic surfactant liquid crystals using synchrotron radiation. *J. Synchrotron Radiat.* **2005**, *12*, 717–720. [CrossRef] [PubMed]

54. Zhao, D.D.; Zhou, W.J.; Li, H.L. Effects of deposition potential and anneal temperature on the hexagonal nanoporous nickel hydroxide films. *Chem. Mater.* **2007**, *19*, 3882–3891. [CrossRef]

55. Raman, N.K.; Anderson, M.T.; Brinker, C.J. Template-based approaches to the preparation of amorphous, nanoporous silicas. *Chem. Mater.* **1996**, *8*, 1682–1701. [CrossRef]

56. Dale, J. Reduction of symmetrical 1,2- and 1,3-diketones with sodium borohydride, and separation of diastereoisomeric 1,2- and 1,3-diols by means of borate complexes. *J. Chem. Soc.* **1961**, 910–922. [CrossRef]

nanomaterials

MDPI

Article

Improvement of Image Sticking in Liquid Crystal Display Doped with γ-Fe₂O₃ Nanoparticles

Wenjiang Ye [1], Rui Yuan [1], Yayu Dai [1], Lin Gao [1], Ze Pang [1], Jiliang Zhu [1], Xiangshen Meng [2], Zhenghong He [2], Jian Li [2], Minglei Cai [3,4], Xiaoyan Wang [3,4] and Hongyu Xing [1,*]

[1] School of Sciences, Hebei University of Technology, Tianjin 300401, China; wenjiang_ye@hebut.edu.cn (W.Y.); rui_yuanedu@163.com (R.Y.); 15102225686@163.com (Y.D.); 15983332112@163.com (L.G.); pangze111@163.com (Z.P.); zjl-656969@163.com (J.Z.)
[2] School of Physical Science and Technology, Southwest University, Chongqing 400715, China; 15552267851@163.com (X.M.); hezhenho@swu.edu.cn (Z.H.); aizhong@swu.edu.cn (J.L.)
[3] Hebei Jiya Electronics Co. Ltd., Shijiazhuang 050071, China
[4] Hebei Provincial Research Center of LCD Engineering Technology, Shijiazhuang 050071, China; cml@jiyalcd.com (M.C.); wwxxy@jiyalcd.com (X.W.)
[*] Correspondence: 2004041@hebut.edu.cn; Tel.: +86-22-6043-5632

Received: 25 October 2017; Accepted: 19 December 2017; Published: 24 December 2017

Abstract: Image sticking in thin film transistor-liquid crystal displays (TFT-LCD) is related to the dielectric property of liquid crystal (LC) material. Low threshold value TFT LC materials have a weak stability and the free ions in them will be increased because of their own decomposition. In this study, the property of TFT LC material MAT-09-1284 doped with γ-Fe₂O₃ nanoparticles was investigated. The capacitances of parallel-aligned nematic LC cells and vertically aligned nematic LC cells with different doping concentrations were measured at different temperatures and frequencies. The dielectric constants perpendicular and parallel to long axis of the LC molecules $\varepsilon_\perp$ and $\varepsilon_{//}$, as well as the dielectric anisotropy $\Delta\varepsilon$, were obtained. The dynamic responses and the direct current threshold voltages in parallel-aligned nematic LC cells for different doping concentrations were also measured. Although the dielectric anisotropy $\Delta\varepsilon$ decreased gradually with increasing temperature and frequency at the certain frequency and temperature in LC state for each concentration, the doping concentration of γ-Fe₂O₃ nanoparticles less than or equal to 0.145 wt % should be selected for maintaining dynamic response and decreasing free ions. This study has some guiding significance for improving the image sticking in TFT-LCD.

Keywords: image sticking; dielectric property; liquid crystal; γ-Fe₂O₃ nanoparticles; capacitance model; dynamic response; threshold voltage

1. Introduction

The application of liquid crystal (LC) materials has widely permeated the display and non-display fields. LC display is realized by controlling the molecular orientation of LCs under an external voltage [1]. In the non-display field, LCs can modulate the phase through LC devices. The adjustment effect is improved by using different device structures and material designs, such as an in-plane switching mode [2], fringe field switching mode [3–5], multi-domain structure [6–8], polymer-stabilized LC [9–13], and polymer network LC [14–16]. Another method is doping nanoparticles into LC materials [17–28]. Nanoparticles doped into LC materials [26–28] can increase the dielectric anisotropy of them to decrease the threshold voltage. However, the low threshold value LC materials are unsuitable for thin film transistor-liquid crystal displays (TFT-LCD) because image sticking always occurs [29]. Image sticking plays an important role in the image quality of liquid crystal displays (LCDs), by which the previous pattern is still visible when the next pattern is addressed. Some papers have studied the improvement of the image sticking through the structure and the composition of TFT-LCD alignment film materials [30–33]. To reduce image sticking, reducing the transient bound ions on the alignment layers in LC materials is a feasible method. Low threshold value TFT LC materials have weak stability and the free ions in them will be increased because of their own decomposition, which leads to the increase of transient bound ions. So high threshold value TFT LC materials are usually chosen, such as MAT-09-1284 (Merck, Darmstadt, Germany). As a type of magnetic nanoparticle, γ-Fe$_2$O$_3$ nanoparticles doped into LC materials can also change the LC molecular orientation and thereby cause a magneto-optic effect [34]. On the other hand, the dielectric anisotropy of the liquid crystal material has a certain influence on the image sticking. Therefore, the dielectric properties are subject to further investigation. In this study, we investigated the dielectric property of LC material MAT-09-1284 doped with γ-Fe$_2$O$_3$ nanoparticles to explore how to improve image sticking in TFT-LCD.

The dielectric property of LC material mainly refers to the change of dielectric constants. In general, the factors that affect the LC dielectric constants include the LC molecular structure, LC blending, LC doping, temperature, and external voltage frequency. Once the LC composition is determined, only temperature and the external voltage frequency affect the dielectric constants of LCs. Given the electric anisotropy of the dielectric material, the dielectric constants of LC materials are described by $\varepsilon_\perp$ and $\varepsilon_{//}$, which represent perpendicular and parallel orientations, respectively, relative to the long axis of LC molecules. Consequently, the dielectric property of LC material MAT-09-1284 doped with γ-Fe$_2$O$_3$ nanoparticles can be investigated by measuring their dielectric constants.

Normally, the dielectric constants $\varepsilon_\perp$ and $\varepsilon_{//}$ can be obtained by measuring the capacitance of parallel-aligned nematic (PAN) cell under low voltage (less than the threshold voltage) and high voltage (greater than the saturation voltage) [35], respectively. However, great errors are highly likely under this method. First, the alignment layer on the glass substrate surface influences the measured capacitance of the LC cell, and its capacitance should also be considered [36]. Second, given the strong anchoring of the alignment layer to LC molecules in a PAN cell, guaranteeing the orientation of all LC molecules along the electric field is highly difficult even when a high voltage is applied to LC cells. Thus, measuring the dielectric constant $\varepsilon_{//}$ as mentioned above causes an important problem. In this regard, the LC cell capacitance can be transformed into the LC layer capacitance by using the LC cell capacitance model. The dielectric constants $\varepsilon_\perp$ and $\varepsilon_{//}$ can be measured by the dual-cell method [37], namely, they are obtained by the LC layer capacitance of PAN cell for low voltage (less than the threshold voltage) and vertically aligned nematic (VAN) cell for high voltage, respectively.

Besides the dielectric property, the doped nanoparticles also affect other physical properties of the LC host, e.g., viscosity. For this purpose, the dynamic responses of LC material MAT-09-1284 doped with γ-Fe$_2$O$_3$ nanoparticles in PAN cells for different concentrations were also conducted and analyzed.

2. Preparation of LC Material and LC cell

2.1. Preparation of LC Material Doped with Nanomaterials

γ-Fe$_2$O$_3$ nanoparticles were prepared by the chemically induced transition method [38–40]. First, precursor FeOOH/Mg(OH)$_2$ was synthesized by the chemical co-precipitation method. Subsequently, the resultant hydroxide precursor FeOOH/Mg(OH)$_2$ was treated in the liquid phase with FeCl$_2$ solution. During treatment, the Mg(OH)$_2$ compound dissolved and the FeOOH dehydrated and transformed into γ-Fe$_2$O$_3$ nanoparticles. LC material MAT-09-1284 was purchased from Merck. It was a mixture of various LC monomers. The clearing point of the pure LC mixture without doping the γ-Fe$_2$O$_3$ nanoparticles was 80.5 °C, the perpendicular and parallel dielectric constants $\varepsilon_\perp$ and $\varepsilon_{//}$ at the temperature 25 °C are 5.2 and 2.6, respectively, and the dielectric anisotropy $\Delta\varepsilon$ at the temperature 25 °C is 2.6. γ-Fe$_2$O$_3$ nanoparticle is an inorganic material, and LC material MAT-09-1284 is an organic material; thus, the components are not mutually soluble. Only by coating γ-Fe$_2$O$_3$ nanoparticles with oleinic acid can they blend well together at different concentrations.

2.2. Preparation of LC Cell

The schematic of a LC cell is shown in Figure 1. The LC cell comprises two etched indium tin oxide (ITO) glass substrates, each of which is coated with polyimide (PI) and sealed by a border adhesive. The LC cells used in this experiment include PAN and VAN cells, which were prepared with the following steps: ITO glass cleaning → drying → etching → cleaning → drying → PI coating → drying → rubbing → cleaning → drying → border sealing. As the alignment layer, the PI layer can guarantee the LC molecules homogeneous alignment, which is crucial to LC devices. The relative dielectric constants of the parallel and vertical PIs are all 3.1, and the pre-tilt angles on these two glass substrates in PAN and VAN cells are 1° and 89°, respectively. The effective electric field is a 1-cm-diameter circle which was etched in the ITO glass substrate.

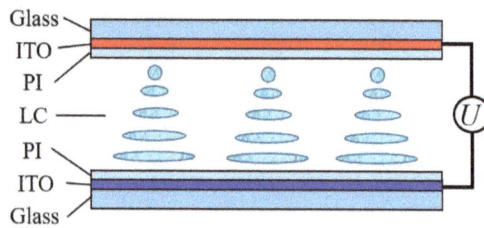

Figure 1. Schematic of LC cell capacitance model.

2.3. LC Cell Capacitance Model

A LC cell can be viewed as a capacitor from its configuration (Figure 1). The cell possesses a capacitance composed of three parts, namely, the upper and lower PI alignment layer capacitances and the LC layer capacitance [36,41]. LC cell capacitor is equivalent to a series of three-layer capacitors. The measured capacitance actually corresponds to the capacitance of LC cell, and the external voltage is also applied to LC cell. However, the LC layer capacitance determines the dielectric constants of LC material. This observation indicates that LC cell capacitance must be transformed into LC layer capacitance by using the LC cell capacitance model.

LC cell capacitance C is given by

$$\frac{1}{C} = \frac{1}{C_1} + \frac{1}{C_{LC}} + \frac{1}{C_2} \tag{1}$$

where $C_1 = S\varepsilon_0\varepsilon_1/L_1$ and $C_2 = S\varepsilon_0\varepsilon_2/L_2$ correspond to the capacitances of upper and lower PI layers, respectively; C_{LC} is the capacitance of LC layer; S is the electrode area; ε_1 and L_1 are the relative dielectric constant and the thickness of the upper PI layer, respectively; and ε_2 and L_2 are the relative dielectric constant and the thickness of the lower PI layer, respectively. The expression of C_{LC} can be easily derived as

$$C_{LC} = \frac{CC_1C_2}{C_1C_2 - CC_2 - CC_1} \tag{2}$$

The external voltage U_{LC} applied to the LC layer is expressed as

$$U_{LC} = \frac{CU}{C_{LC}} \tag{3}$$

where U is the external voltage applied to LC cell.

If the influence of the pre-tilt angles θ_{PAN} and θ_{VAN} on the substrate surface in PAN and VAN cells is considered, then the dielectric constants $\varepsilon_\perp$ and $\varepsilon_{//}$ of LC material satisfy the equation

$$\begin{cases} \dfrac{1}{C_{LC-PAN}} = \dfrac{L_{LC-PAN}}{S\varepsilon_0\left(\varepsilon_\perp + \Delta\varepsilon \sin^2\theta_{PAN}\right)} \\ \dfrac{1}{C_{LC-VAN}} = \dfrac{L_{LC-VAN}}{S\varepsilon_0\left(\varepsilon_\perp + \Delta\varepsilon \sin^2\theta_{VAN}\right)} \end{cases} \tag{4}$$

where L_{LC-PAN} and L_{LC-VAN} are the thicknesses of LC layer in PAN and VAN cells, respectively; $\Delta\varepsilon = \varepsilon_{//} - \varepsilon_\perp$ is the dielectric anisotropy of LC material; and ε_0 is the vacuum dielectric constant.

3. Experiment

The instrument used to measure the LC cell capacitance was the precision LCR meter E4980A (Agilent, Palo Alto, CA, USA). The experimental configuration was shown in Figure 2A. Certain temperature was ensured in the measurement by mounting the LC cell on a hot stage LTS350 (Linkam, Surrey, UK) regulated by a hot controller TP94 (Linkam). At the same time, the effects of the lead wires and the alligator clips connected to the test fixture were eliminated by minimizing the length of the lead wires.

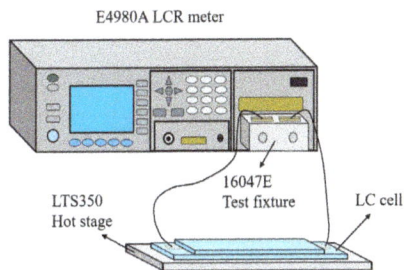

Figure 2. Experimental configuration for measuring the LC cell capacitance.

First, the LC material MAT-09-1284 doped with γ-Fe_2O_3 nanoparticles was used to fill the PAN and VAN cells, which were sealed by the ultraviolet (UV) sealing adhesive. During the UV curing process, the polarizer was attached on the glass substrate above the LC layer to protect the LC from UV light. Next, metal pins were added to both LC cell substrates through conductive adhesive, and the LC cell was fixed on a hot stage by high-temperature-resistant adhesive tape. Then, the test fixture 16047E (Agilent, Palo Alto, CA, USA) was connected to the precision LCR meter E4980A (Agilent). After setting up the measurement conditions of the precision LCR meter, the open/short correction function was applied to acquire further precise data. Finally, the capacitance data with different doping concentrations were recorded at different temperatures and frequencies under external voltages from 0.1 to 20 V and used to plot capacitance-voltage curves. Through the LC cell capacitance model, the dielectric constants and dielectric anisotropy could be obtained.

The LC layer capacitances of PAN and VAN cells under different voltages were obtained by accurately measuring the thicknesses of the LC layer (cell gap) and those of the upper and lower PI alignment layers. Through the double-beam UV and visible spectrophotometer UV-9000S (Metash, Shanghai, China), the average values of the cell gap of PAN and VAN cells were 3.95 and 4.00 µm, respectively. Given different PIs used in manufacturing PAN and VAN cells, the thicknesses of these two PI alignment layers differed. With the aid of the non-contact surface profilometer Contor GT-K (Bruker, Karlsruhe, Germany), the average thicknesses of the PI layers in PAN and VAN cells were 50 and 15 nm, respectively.

3.1. Influence of Temperature on the Dielectric Property

The dielectric constants of LC materials were all known to be obviously influenced by temperature. Only when the temperature was within a certain range would the LC materials be in the LC state. The influence of temperature on the dielectric constants of the LC material MAT-09-1284 doped with γ-Fe_2O_3 nanoparticles was investigated under different concentrations. We measured the capacitances of PAN and VAN cells from the temperature of 20 to 100 °C by adjusting the hot controller, as shown in Figures 3 and 4. The concentrations of doped γ-Fe_2O_3 nanoparticles were (a) 0.0; (b) 0.02; (c) 0.048; (d) 0.145; (e) 0.515; (f) 0.984; and (g) 2.6 wt %. The frequency of the external voltage was 1 kHz.

From the microscopic viewpoint, the dielectric constants $\varepsilon_\perp$ and $\varepsilon_{//}$ of the LC material MAT-09-1284 doped with γ-Fe_2O_3 nanoparticles are related to the molecular polarization, the order parameter, the angle between the permanent dipole moment and the molecular long axis, and the magnetization of γ-Fe_2O_3 nanoparticles excited by the electric field, etc. This is a complicated change which can be reflected by the capacitance of LC cell. When the temperature was certain and the voltage was lower than the threshold voltage, the capacitance of PAN cell was only related to the molecular polarization and the magnetization. The general tendency of the capacitance was initially decreased and then increased before the temperature reached the clearing point as the nanoparticle concentration increased, as shown in Figure 5a. The clearing point decreased with the increase in the concentration of doped γ-Fe_2O_3 nanoparticles (Figure 3). When the doping concentration reached 2.6 wt %, the capacitance achieved a small change, which indicated that the doped γ-Fe_2O_3 nanoparticles induced the LC material MAT-09-1284 to assume an almost isotropic state. This result revealed that γ-Fe_2O_3 nanoparticles doped into LC materials changed the LC molecular orientation by their magnetizations. In VAN cells, the capacitance decreased monotonously and remained unchanged with different voltages for the doping concentrations less than 2.6 wt %, under which around a dozen changed in the capacitance (Figure 4). The corresponding changes in the dielectric constant $\varepsilon_{//}$ were shown in Figure 5b. This result illustrated that the LC material MAT-09-1284 doped with γ-Fe_2O_3 nanoparticles remained in the LC state, and the dielectric anisotropy $\Delta\varepsilon$ decreased with the increase in temperature. When the doping concentration was less than or equal to 0.145 wt % and the temperature was below the clearing point, the dielectric anisotropy changed slightly, especially at 0.048 wt % (Figure 5c). This result revealed that such concentration did not affect the dielectric anisotropy $\Delta\varepsilon$ of the LC MAT-09-1284 even though the dielectric constants $\varepsilon_\perp$ and $\varepsilon_{//}$ have some differences.

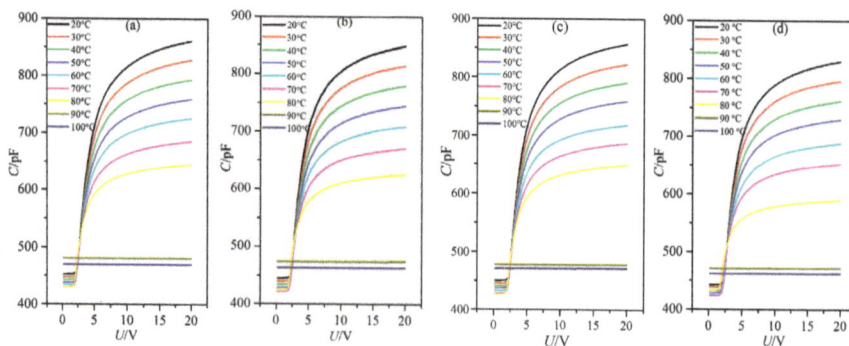

Figure 3. *Cont.*

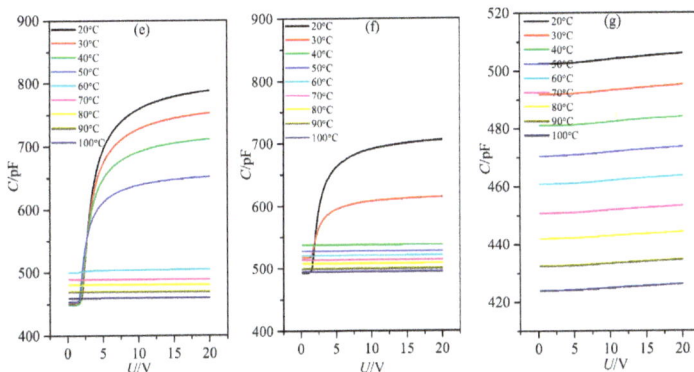

Figure 3. Liquid crystal (LC) cell capacitance versus voltage with frequency of 1 kHz for parallel-aligned nematic (PAN) cell under different temperatures and doped γ-Fe_2O_3 nanoparticle concentrations of (**a**) 0.0; (**b**) 0.02; (**c**) 0.048; (**d**) 0.145; (**e**) 0.515; (**f**) 0.984; and (**g**) 2.6 wt %.

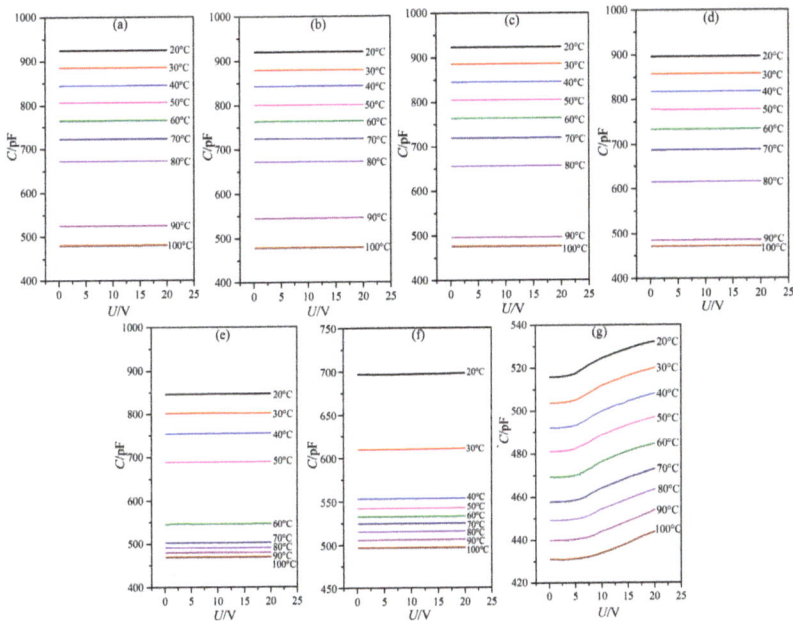

Figure 4. LC cell capacitance versus voltage with frequency of 1 kHz for vertically aligned nematic (VAN) cell under different temperatures and doped γ-Fe_2O_3 nanoparticle concentrations of (**a**) 0.0; (**b**) 0.02; (**c**) 0.048; (**d**) 0.145; (**e**) 0.515; (**f**) 0.984; and (**g**) 2.6 wt %.

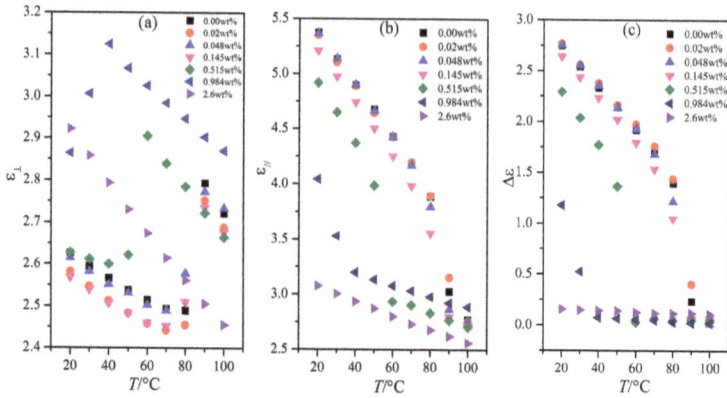

Figure 5. Dependence of the dielectric constants (a) $\varepsilon_\perp$ and (b) $\varepsilon_{//}$ and the dielectric anisotropy (c) $\Delta\varepsilon$ on temperature for the LC material MAT-09-1284 doped with γ-Fe$_2$O$_3$ nanoparticles of different concentrations.

3.2. Influence of External Voltage Frequency on the Dielectric Property

Aside from temperature, the external voltage frequency can also affect the LC dielectric constants. The temperature of the hot stage was stabilized to 25 °C, and the external voltage frequency was then changed by adjusting the precision LCR meter from 20 Hz to 2 kHz. The capacitances of PAN cell with an external voltage of 1 V (less than the threshold voltage in Figure 3) and VAN cell with an external voltage of 15 V (aligned well along the direction of the electric field) for the LC material MAT-09-1284 doped with γ-Fe$_2$O$_3$ nanoparticles of different concentrations were measured (Figures 6 and 7).

For PAN and VAN cells, the capacitance variations with the external voltage frequency for different doping concentrations were extremely small (only several pF) within the frequency measurement range, except for the relatively low frequency of less than 100 Hz (Figures 6 and 7). This result inevitably resulted in the small variations of the dielectric constants $\varepsilon_\perp$ and $\varepsilon_{//}$ and the dielectric anisotropy $\Delta\varepsilon$ to the frequency, as shown in Figure 8. In these cases, the changes of the dielectric constants were controlled by molecular polarization and magnetization. Although these variation rules on the dielectric constants $\varepsilon_\perp$ and $\varepsilon_{//}$ had some differences, the dielectric anisotropy $\Delta\varepsilon$ was decreased gradually with increased doping concentration for a certain frequency. When the doping concentration was less than 0.145 wt %, the difference was relatively small.

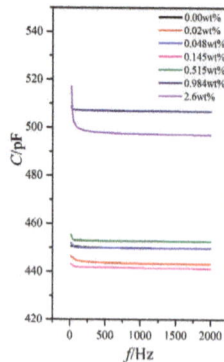

Figure 6. PAN cell capacitance versus frequency with different doped γ-Fe$_2$O$_3$ nanoparticle concentrations under the external voltage of 1 V and temperature of 25 °C.

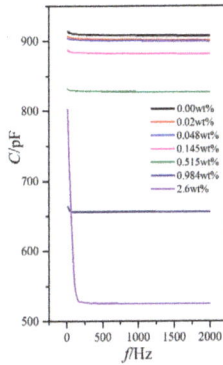

Figure 7. VAN cell capacitance versus frequency with different doped γ-Fe$_2$O$_3$ nanoparticle concentrations under the external voltage of 15 V and temperature 25 °C.

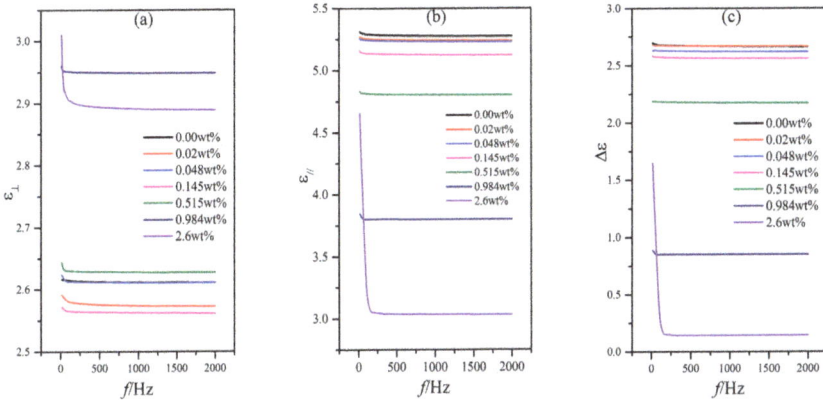

Figure 8. Dependence of the dielectric constants (**a**) $\varepsilon_\perp$; (**b**) $\varepsilon_{//}$ and the dielectric anisotropy (**c**) $\Delta\varepsilon$ on the frequency for the LC material MAT-09-1284 doped with γ-Fe$_2$O$_3$ nanoparticles of different concentrations.

3.3. Dynamic Response and DC Threshold Voltage

Through the preceding analyses on the dielectric property of LC material MAT-09-1284 doped with γ-Fe$_2$O$_3$ nanoparticles, an appropriate doping concentration could be selected to decrease the dielectric anisotropy for improving the image sticking. However, the response time for such doping concentration should be guaranteed. Meanwhile, to explain the cause of improving the image sticking in nature, the behaviors of LC material MAT-09-1284 with different doping concentrations under DC drive should also be analyzed.

Impulse voltage with time 100 ms and amplitude 8 V or a DC voltage from 0 to 20 V was applied to PAN cells. The changes in the intensity of a laser beam (632.8 nm) that passed through the PAN cells were monitored by a detector connected to the oscilloscope (Tektronix MDO3024, Johnston, OH, USA). The temperature was controlled at 20 °C. The curves of normalized transmittance versus time and DC threshold voltage versus doping concentration in PAN cells were shown in Figures 9 and 10, respectively.

The dynamic responses of LC material MAT-09-1284 for different doping concentrations less than or equal to 0.145 wt % were almost identical which could be seen from Figure 9. In addition, it was inevitable that there would be ions in the LC materials, as well as in the LC MAT-09-1284. These ions in PAN cells could be divided into two parts: free ions moving with the electric field and transient bound ions on the alignment layers. When the voltage induced by the bound ions was greater than the threshold voltage, image sticking would occur [42]. The free ions, however, could counteract the ion electric field to reduce the occurrence of image sticking. When a small amount of γ-Fe_2O_3 nanoparticles ($\leq$0.145 wt %) was doped into the LC material MAT-09-1284, the magnetization of nanoparticles excited by electric field might adsorb the free ions, which would result in the increase of the DC threshold voltage, as shown in Figure 10. In this case, the magnetization was ordered. As increasing the doping concentration, the order of magnetization decreased and more free ions moved to the alignment layers to reduce the DC threshold voltage. Therefore, the dynamic response and the DC threshold voltage considered, the doping concentration ($\leq$0.145 wt %) should be selected to improve the image sticking.

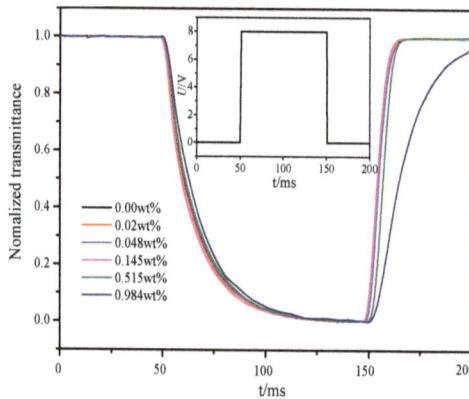

Figure 9. Normalized transmittance versus time of the LC material MAT-09-1284 doped with γ-Fe_2O_3 nanoparticles for different concentrations in PAN cell. The embedded diagram describes the characteristic of the pulse voltage. The pulse width is 100 ms and the amplitude is 8 V.

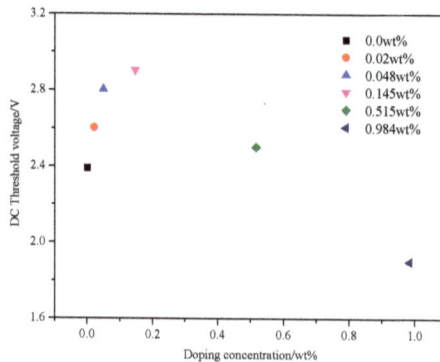

Figure 10. DC threshold voltage versus doping concentration of the LC material MAT-09-1284 doped with γ-Fe_2O_3 nanoparticles in PAN cell.

3.4. Long Term Stability of the LC Mixture

Most nanoparticle suspended LC systems suffer from long term stability issues. For investigating the long term stability of LC material MAT-09-1284 doped with γ-Fe_2O_3 nanoparticles of different doping concentrations, we controlled the same experimental environment as before and remeasured the capacitance of PAN cells at different temperatures and the frequency of 1 kHz to compare them with the original data from one month prior, as shown in Figure 11. The concentrations of doped γ-Fe$_2$O$_3$ nanoparticles were (a) 0.0 wt %; (b) 0.02 wt %; (c) 0.048 wt %; (d) 0.145 wt %; (e) 0.515 wt %; and (f) 0.984 wt %.

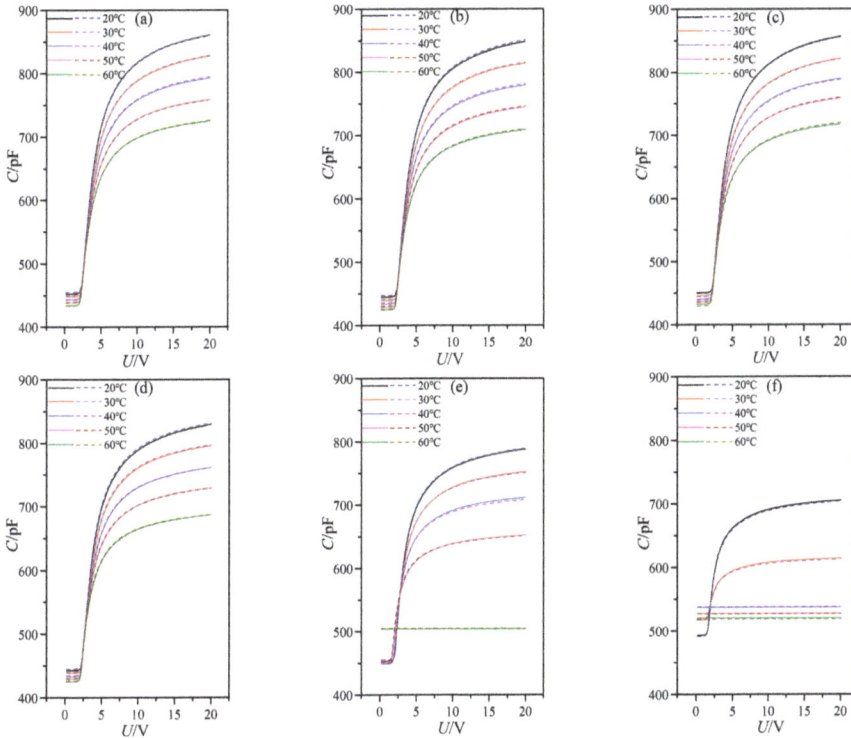

Figure 11. LC cell capacitance versus voltage with frequency of 1 kHz for PAN cell under different temperatures and doped γ-Fe$_2$O$_3$ nanoparticle concentrations of (a) 0.0 wt %; (b) 0.02 wt %; (c) 0.048 wt %; (d) 0.145 wt %; (e) 0.515 wt %; and (f) 0.984 wt %. Solid lines for the original data and dashed lines for the data a month later.

These curves of capacitance versus voltage were coincident for two independent measurements. It showed that γ-Fe$_2$O$_3$ nanoparticles used in experiment blended well with the LC material MAT-09-1284 and they did not aggregate in the LC host as the time increased. In other words, the LC material MAT-09-1284 doped with γ-Fe$_2$O$_3$ nanoparticles of different doping concentrations possessed long term stability.

4. Conclusions

In this study, through investigating the effect of temperature and frequency on the dielectric property of LC material MAT-09-1284 doped with γ-Fe$_2$O$_3$ nanoparticles of different doping concentrations, the dynamic response and the DC threshold voltage in PAN cell, an improvement of image sticking in TFT-LCD was proposed. In addition to selecting the high threshold value TFT LC materials, doping γ-Fe$_2$O$_3$ nanoparticles into the LC materials could also improve the image sticking because of their magnetization induced by the electric field. Compared to the pure LC material

MAT-09-1284, a doping concentration less than or equal to 0.145 wt % of γ-Fe$_2$O$_3$ nanoparticles could be chosen. The best doping concentration was 0.145 wt %, with which the long term stability of LC mixture could be maintained. This study has some guiding significance for improving image sticking in TFT-LCD.

Acknowledgments: This research is supported by the National Natural Science Foundation of China (Grant Nos. 11374087, 11504080), the Natural Science Foundation of Hebei Province (Grant Nos. A2014202123, A2017202004), the Research Project of Hebei Education Department (Grant No. QN2014130), and the Key Subject Construction Project of Hebei Province University.

Author Contributions: Wenjiang Ye and Hongyu Xing conceived and designed the experiments; Yayu Dai, Lin Gao, and Ze Pang performed the experiments; Rui Yuan, Jiliang Zhu, and Wenjiang Ye analyzed the data; Minglei Cai and Xiaoyan Wang contributed the LC cells; Xiangshen Meng, Zhenghong He, and Jian Li contributed the LC material MAT-09-1284 and nanoparticles γ-Fe$_2$O$_3$; Rui Yuan, Wenjiang Ye, and Hongyu Xing wrote the paper.

Conflicts of Interest: The authors declare no conflict of interest.

References

1. Yang, D.K.; Wu, S.T. *Fundamentals of Liquid Crystal Devices*, 1st ed.; John Wiley & Sons, Ltd.: Chichester, UK, 2006.
2. Sun, Y.B.; Zhao, Y.L.; Li, Y.F.; Li, P.; Ma, H.M. Optimisation of in-plane-switching blue-phase liquid crystal display. *Liq. Cryst.* **2014**, *41*, 717–720. [CrossRef]
3. Lee, S.H.; Lee, S.L.; Kim, H.Y. Electro-optic characteristics and switching principle of a nematic liquid crystal cell controlled by fringe-field switching. *Appl. Phy. Lett.* **1998**, *73*. [CrossRef]
4. Yun, H.J.; Jo, M.H.; Jang, I.W.; Lee, S.H.; Ahn, S.H.; Hur, H.J. Achieving high light efficiency and fast response time in fringe field switching mode using a liquid crystal with negative dielectric anisotropy. *Liq. Cryst.* **2012**, *39*. [CrossRef]
5. Chen, Y.; Luo, Z.; Peng, F.; Wu, S.T. Fringe-field switching with a negative dielectric anisotropy liquid crystal. *J. Disp. Technol.* **2013**, *9*, 74–77. [CrossRef]
6. Ma, J.; Yang, Y.C.; Zheng, Z.; Shi, J.; Cao, W. A multi-domain vertical alignment liquid crystal display to improve the V–T property. *Displays* **2009**, *30*, 185–189. [CrossRef]
7. Lee, Y.J.; Kim, Y.K.; Jo, S.I.; Bae, K.S.; Choi, B.D.; Kim, J.H.; Yu, C.J. Fast vertical alignment mode with continuous multi-domains for a liquid crystal display. *Opt. Express* **2009**, *17*, 23417–23422. [CrossRef] [PubMed]
8. Park, S.; Lim, S.; Choi, Y.; Jeong, K.; Lee, M.; Chang, H.; Kim, H.; Lee, S. Multi-domain vertical alignment liquid crystal displays with ink-jet printed protrusions. *Liq. Cryst.* **2012**, *39*, 501–507. [CrossRef]
9. Kikuchi, H.; Yokota, M.; Hisakado, Y.; Yang, H.; Kajiyama, T. Polymer-stabilized liquid crystal blue phases. *Nat. Mater.* **2002**, *1*, 64–68. [CrossRef] [PubMed]
10. Lu, S.Y.; Chien, L.C. A polymer-stabilized single-layer color cholesteric liquid crystal display with anisotropic reflection. *Appl. Phy. Lett.* **2007**, *91*, 131119. [CrossRef]
11. Xu, X.W.; Zhang, X.W.; Luo, D.; Dai, H.T. Low voltage polymer-stabilized blue phase liquid crystal reflective display by doping ferroelectric nanoparticles. *Opt. Express* **2015**, *23*, 32267–32273. [CrossRef] [PubMed]
12. Zhou, X.C.; Qin, G.K.; Yang, D.K. Single-cell gap polymer-stabilized fringe-field switching transflective liquid crystal display. *Opt. Lett.* **2016**, *41*, 257–260. [CrossRef] [PubMed]
13. You, R.K.; Choi, Y.E.; Wen, P.; Lee, B.H.; Kim, J.C.; Lee, M.H.; Jeong, K.U.; Lee, S.H. Polymer stabilized vertical alignment liquid crystal display: Effect of monomer structures and their stabilizing characteristics. *J. Phys. D Appl. Phys.* **2016**, *49*, 165501.
14. Hasebe, H.; Takatsu, H.; Iimura, Y.; Kobayashi, S. Effect of polymer network made of liquid crystalline diacrylate on characteristics of liquid crystal display device. *Jpn. J. Appl. Phys.* **1994**, *33*, 6245–6248. [CrossRef]
15. White, T.J.; Broer, D.J. Programmable and adaptive mechanics with liquid crystal polymer networks and elastomers. *Nat. Mater.* **2015**, *14*, 1087–1098. [CrossRef] [PubMed]
16. Xu, D.; Peng, F.; Chen, H.W.; Wu, S.T. Low voltage polymer network liquid crystal for infrared spatial light modulators. *Opt. Express* **2015**, *23*, 2361–2368.

17. Algorri, J.F.; García-Cámara, B.; Urruchi, V.; Sánchez-Pena, J.M. High-sensitivity Fabry-Pérot temperature sensor based on liquid crystal doped with nanoparticles. *IEEE Photonics Technol. Lett.* **2015**, *27*, 292–295. [CrossRef]

18. Ganguly, P.; Kumar, A.; Muralidhar, K.; Biradar, A.M. Nanoparticles induced multiferroicity in liquid crystal. *Appl. Phy. Lett.* **2016**, *108*, 182905. [CrossRef]

19. Yadav, S.P.; Manohar, R.; Singh, S. Effect of TiO_2 nanoparticles dispersion on ionic behaviour in nematic liquid crystal. *Liq. Cryst.* **2015**, *42*, 1095–1101. [CrossRef]

20. Mishra, K.G.; Dubey, S.K.; Mani, S.A.; Pradhan, M.S. Comparative study of nanoparticles doped in liquid crystal polymer system. *J. Mol. Liq.* **2016**, *224*, 668–671. [CrossRef]

21. He, W.L.; Zhang, W.K.; Xu, H.; Li, L.H.; Yang, Z.; Cao, H.; Wang, D.; Zheng, Z.G.; Yang, H. Preparation and optical properties of Fe_3O_4 nanoparticles-doped blue phase liquid crystal. *Phys. Chem. Chem. Phys.* **2016**, *18*, 29028–29032. [CrossRef] [PubMed]

22. Prodanov, M.F.; Buluy, O.G.; Popova, E.V.; Gamzaeva, S.A.; Reznikov, Y.O.; Vashchenko, V.V. Magnetic actuation of a thermodynamically stable colloid of ferromagnetic nanoparticles in a liquid crystal. *Soft Matter* **2016**, *12*, 6601–6609. [CrossRef] [PubMed]

23. Chen, W.T.; Chen, P.S.; Chao, C.Y. Effect of doped insulating nanoparticles on the electro-optical characteristics of nematic liquid crystals. *Jpn. J. Appl. Phys.* **2009**, *48*, 015006. [CrossRef]

24. Singh, U.B.; Dhar, R.; Dabrowski, R.; Pandey, M.B. Enhanced electro-optical properties of a nematic liquid crystals in presence of $BaTiO_3$ nanoparticles. *Liq. Cryst.* **2014**, *41*, 953–959. [CrossRef]

25. Al-Zangana, S.; Turner, M.; Dierking, I. A comparison between size dependent paraelectric and ferroelectric $BaTiO_3$ nanoparticle doped nematic and ferroelectric liquid crystals. *J. Appl. Phys.* **2017**, *121*, 85105. [CrossRef]

26. Kobayashi, S.; Miyama, T.; Nishida, N.; Sakai, Y.; Shiraki, H.; Shiraishi, Y.; Toshima, N. Dielectric spectroscopy of metal nanoparticle doped liquid crystal displays exhibiting frequency modulation response. *J. Disp. Technol.* **2006**, *2*, 121–129. [CrossRef]

27. Scolari, L.; Gauza, S.; Xianyu, H.Q.; Zhai, L.; Eskildsen, L.; Alkeskjold, T.T.; Wu, S.T.; Bjarklev, A. Frequency tunability of solid-core photonic crystal fibers filled with nanoparticle-doped liquid crystals. *Opt. Express* **2009**, *17*, 3754–3764. [CrossRef] [PubMed]

28. Shiraishi, Y.; Toshima, N.; Maeda, K.; Yoshikawa, H.; Xu, J.; Kobayashi, S. Frequency modulation response of a liquid-crystal electro-optic device doped with nanoparticles. *Appl. Phys. Lett.* **2002**, *81*, 2845. [CrossRef]

29. Stojmenovik, G.; Neyts, K.; Vermael, S.; Verschueren, A.R.M.; Asselt, R. The influence of the driving voltage and ion concentration on the lateral ion transport in nematic liquid crystal displays. *Jpn. J. Appl. Phys.* **2005**, *44*, 6190. [CrossRef]

30. Park, Y.; Kim, S.; Lee, E. A study on reducing image-sticking artifacts in wide-screen TFT-LCD monitors. *J. Soc. Inf. Disp.* **2007**, *15*, 969–973. [CrossRef]

31. Mizusaki, M.; Nakanishi, Y. Improvement of image sticking on liquid crystal displays with polymer layers produced from mixed monomers. *Liq. Cryst.* **2016**, *43*, 704–710. [CrossRef]

32. Sang, G.L.; Hong, S.H.; Hwang, Y.J.; Dong, M.S. P-162: A Study on Image Sticking of Photo Reactive Polyimide Alignment Layer for LCD. *SID Symp. Dig. Tech. Pap.* **2008**, *39*, 1808–1810.

33. Lee, T.R.; Jin, H.K.; Lee, S.H.; Hong, K.B.; Jun, M.C. Investigation on newly designed low resistivity polyimide-type alignment layer for reducing DC image sticking of in-plane switching liquid crystal display. *Liq. Cryst.* **2016**, *44*, 738–747. [CrossRef]

34. Domracheva, N.E.; Pyataev, A.V.; Manapov, R.A.; Gruzdev, M.S. Magnetic resonance and mössbauer studies of superparamagnetic γ-Fe_2O_3 Nanoparticles Encapsulated into Liquid-Crystalline Poly (propylene imine) Dendrimers. *Chemphyschem* **2011**, *12*, 3009–3019. [CrossRef] [PubMed]

35. Lai, C.W.; Ting, T.L.; Hsu, W.H. Dielectric constant measurement of polyimide and liquid crystal at low frequency. *SID Symp. Dig. Tech. Pap.* **2016**, *47*, 1679–1680. [CrossRef]

36. Jiao, M.Z.; Ge, Z.B.; Song, Q.; Wu, S.T. Alignment layer effects on thin liquid crystal cells. *Appl. Phy. Lett.* **2008**, *92*, 061102. [CrossRef]

37. Wu, S.T.; Wu, C.S. Experimental confirmation of the Osipov-Terentjev theory on the viscosity of nematic liquid crystals. *Phys. Rew. A* **1990**, *42*, 2219. [CrossRef]

38. Wen, B.C.; Li, J.; Lin, Y.Q.; Liu, X.D.; Fu, J.; Miao, H.; Zhang, Q.M. A novel preparation method for γ-Fe_2O_3 nanoparticles and their characterization. *Mater. Chem. Phys.* **2011**, *128*, 35–38. [CrossRef]

39. Chen, Y.S.; Chen, Q.; Mao, H.; Zhang, T.; Qiu, X.Y.; Lin, Y.Q.; Li, J. Preparation of magnetic nanoparticles via chemically induced transition: Dependence of components and magnetization on the concentration of treating solution used. *Nanomater. Nanotechnol.* **2017**, *7*, 1–9. [CrossRef]

40. Zhang, T.; Meng, X.S.; He, Z.H.; Lin, Y.Q.; Liu, X.D.; Li, D.C.; Li, J.; Qiu, X.Y. Preparation of magnetic nanoparticles via chemically induced transition: Role of treating solution's temperature. *Nanomaterials* **2017**, *7*, 220. [CrossRef] [PubMed]

41. Seiberle, H.; Schadt, M. LC-Conductivity and Cell Parameters; Their Influence on Twisted Nematic and Supertwisted Nematic Liquid Crystal Displays. *Mol. Cryst. Liq. Cryst.* **1994**, *239*, 229–244. [CrossRef]

42. Neyts, K.; Vermael, S.; Desimpel, C.; Stojmenovik, G.; Verschueren, A.R.M.; Boer, D.K.G.; Snijkers, R.; Machiels, P.; Brandenburg, A. Lateral ion transport in nematic liquid-crystal devices. *J. Appl. Phys.* **2003**, *94*, 3891–3896. [CrossRef]

nanomaterials

MDPI

Article

Kinetics of Ion-Capturing/Ion-Releasing Processes in Liquid Crystal Devices Utilizing Contaminated Nanoparticles and Alignment Films

Yuriy Garbovskiy

UCCS BioFrontiers Center and Department of Physics, University of Colorado Colorado Springs,
Colorado Springs, CO 80918, USA; ygarbovs@uccs.edu or ygarbovskiy@gmail.com; Tel.: +1-719-255-3123

Received: 4 January 2018; Accepted: 19 January 2018; Published: 23 January 2018

Abstract: Various types of nanomaterials and alignment layers are considered major components of the next generation of advanced liquid crystal devices. While the steady-state properties of ion-capturing/ion-releasing processes in liquid crystals doped with nanoparticles and sandwiched between alignment films are relatively well understood, the kinetics of these phenomena remains practically unexplored. In this paper, the time dependence of ion-capturing/ion-releasing processes in liquid crystal cells utilizing contaminated nanoparticles and alignment layers is analyzed. The ionic contamination of both nanodopants and alignment films governs the switching between ion-capturing and ion-releasing regimes. The time dependence (both monotonous and non-monotonous) of these processes is characterized by time constants originated from the presence of nanoparticles and films, respectively. These time constants depend on the ion adsorption/ion desorption parameters and can be tuned by changing the concentration of nanoparticles, their size, and the cell thickness.

Keywords: liquid crystals; ions; contaminated nanoparticles; kinetics; ion-capturing films; ion trapping; adsorption/desorption

---

1. Introduction

Ions in liquid crystals can affect the performance of devices utilizing these materials in different ways. In the majority of cases, liquid crystal devices (displays (LCD), tunable wave plates and variable retarders, filters, and lenses) are driven by electric fields reorienting liquid crystal molecules and changing the properties of the device [1,2]. For this type of applications, ions in liquid crystals are very undesirable objects since they can lead to many negative side effects (image sticking, image flickering, reduced voltage holding ratio, and slow response) thus compromising an overall performance of liquid crystal devices [2,3]. That is why the development of new methods to purify liquid crystals from ions is of utmost importance to the LCD industry. There are also an increasing number of applications relying on ions in liquid crystals such as liquid crystal shutters and optical switches utilizing light scattering effects [4–6].

The dispersion of nanomaterials in liquid crystals has recently emerged as a promising way to control the concentration of mobile ions in liquid crystals ([7] and references therein). The most widely used nanomaterials include carbon-based nano-objects [7,8], dielectric [7,9], semiconductor [7,10], magnetic [7,11], metal [7,12], and ferroelectric [7,13,14] nanoparticles. Numerous studies revealed the complex behavior of nanoparticles in liquid crystals and the possibility of several regimes, namely the ion-capturing regime (the purification of liquid crystals from ions), ion-releasing regime (the contamination of liquid crystals with ions), and no change in the concentration of ions [7,15,16]. The ionic contamination of nanomaterials is considered a major factor determining the type of the regime achieved in experiments [15–24].

So far, the effects of nanomaterials on the concentration of mobile ions in liquid crystals were studied under equilibrium conditions mostly. There are very limited number of papers reporting the time dependence of the concentration of ions $n(t)$ in liquid crystals doped with nanomaterials [25,26]. Moreover, the kinetics of ion-capturing/ion-releasing regimes in liquid crystals doped with contaminated nanoparticles was not discussed at all. Alignment layers constitute a major component of practically any liquid crystal device [1,2,27]. Therefore, it is also very important to consider the combined effect of both nanoparticles and alignment layers on the time dependence $n(t)$. The steady-state ion-capturing properties of several types of alignment layers including polymer-based films [28–30], SiO_x-based films [31–33], and films made of graphene [34] have been reported. While in the case of liquid crystal cells utilizing polyimide alignment layers the kinetics of ion adsorption/ion desorption processes causing the ion-capturing effect was studied in several publications [35–39], there are no publications focused on the kinetics of ion-capturing/ion-releasing processes in liquid crystals sandwiched between contaminated alignment layers. In addition, the combined effect of contaminated nanoparticles and alignment layers on the time dependence of ion-capturing/ion-releasing regimes is also not discussed in existing literature. This paper is aimed at analyzing the kinetics of these regimes in liquid crystal cells utilizing nanomaterials and alignment layers contaminated with ions.

2. Theoretical Model and Results

2.1. Contaminated Nanoparticles in Liquid Crystals

Consider liquid crystals doped with nanoparticles. To simplify the discussion, both liquid crystals and contaminated nanoparticles are characterized by the same type of fully ionized ionic species. In this case, the ion adsorption/ion desorption processes change the concentration of mobile ions in liquid crystals according to the rate Equation (1):

$$\frac{dn}{dt} = -k_a^{NP} n_{NP} A_{NP} \sigma_S^{NP} n(1 - \Theta_{NP}) + k_d^{NP} n_{NP} A_{NP} \sigma_S^{NP} \Theta_{NP} \tag{1}$$

where n is the concentration of mobile ions, A_{NP} is the surface area of a single nanoparticle; n_{NP} is the volume concentration of nanoparticles; σ_S^{NP} is the surface density of all adsorption sites on the surface of a single nanoparticle; k_a^{NP} is the adsorption rate constant; and k_d^{NP} is the desorption rate constant; Θ_{NP} is the fractional surface coverage of nanoparticles defined as $\Theta_{NP} = \frac{\sigma_{NP}}{\sigma_S^{NP}}$ (σ_{NP} is the surface density of adsorption sites on the surface of nanoparticles occupied by ions). The first term of Equation (1) accounts for the adsorption of ions onto the surface of nanoparticles, and the second term describes the ion desorption from the surface of nanoparticles. In the steady-state regime ($\frac{dn}{dt} = 0$) Equation (1) reduces to the Langmuir adsorption isotherm [40]. The discussion of the applicability and limitations of this approach to compute the concentration of mobile ions in liquid crystals can be found in recently published papers [41–43].

The conservation law of the total number of ions can be written as Equation (2):

$$n_0 + n_{NP} A_{NP} \sigma_S^{NP} \nu_{NP} = n + n_{NP} A_{NP} \sigma_S^{NP} \Theta_{NP} \tag{2}$$

where n_0 is the initial concentration of ions in liquid crystals, and ν_{NP} is the contamination factor of nanoparticles. The contamination factor of nanoparticles accounts for their possible ionic contamination [15]. It equals the fraction of the adsorption sites on the surface of nanoparticles occupied by ions-contaminants prior to dispersing them in liquid crystals [15].

The kinetics of ion-capturing/ion-releasing processes in liquid crystals doped with contaminated nanoparticles can be computed by solving Equations (1) and (2). These equations can be solved analytically [44–46]. The general analytical solution is very bulky and not easy to analyze [46]. However, in the majority of the reported experimental studies the observed fractional surface coverage is very low, $\Theta_{NP} << 1$ [16,21]. It allows for some simplifications. In the regime of relatively low surface coverage, an analytical solution can be written as Equation (3):

$$n \approx \frac{n_0 + n_{NP}A_{NP}\sigma_S^{NP}\nu_{NP} + n_{NP}A_{NP}\sigma_S^{NP}(K_{NP}n_0 - \nu_{NP})e^{-(k_a^{NP}n_{NP}A_{NP}\sigma_S^{NP}+k_d^{NP})t}}{1 + K_{NP}n_{NP}A_{NP}\sigma_S^{NP}} \tag{3}$$

In real systems, the values of physical parameters characterizing adsorption-desorption processes (σ_S^{NP}, $K_{NP} = \frac{k_a^{NP}}{k_d^{NP}}$) can vary within very broad limits (Table 1). Therefore, in the present study, their values were selected to reasonably represent existing materials.

Table 1. Examples of existing experimental data.

Materials (Liquid Crystals Doped with Nano-Objects)	Physical Parameters	Ref.
Nematic liquid crystals (E44) doped with anatase nanoparticles (TiO$_2$)	$K_{NP} = 10^{-23}$ m^3, $\sigma_S^{NP} = 0.8 \times 10^{18}$ m^{-2}, $\nu_{NP} = 0.00015$, $n_0 = 3 \times 10^{19}$ m^{-3}, $d = 11.5 \pm 0.5$ μm	[16,47]
Nematic liquid crystals (E7) doped with carbon nanotubes	$K_{NP} = 7 \times 10^{-24}$ m^3, $\sigma_S^{NP} = 10^{18}$ m^{-2}, $\nu_{NP} = 0.0000095$, $n_0 = 2.5 \times 10^{18}$ m^{-3}, $d = 11.3$ μm	[16,48]
Liquid crystals (8OCB) doped with graphene	$K_{NP} = 8 \times 10^{-24}$ m^3, $\sigma_S^{NP} = 0.33 \times 10^{18}$ m^{-2}, $\nu_{NP} = 0.0000085$, $n_0 = 2.6 \times 10^{18}$ m^{-3}, $d = 7.0 \pm 0.5$ μm	[16,49]
Nematic liquid crystals (E44) doped with ferroelectric nanoparticles (BaTiO$_3$)	$K_{NP} = 10^{-23}$ m^3, $\sigma_S^{NP} = 5 \times 10^{18}$ m^{-2}, $\nu_{NP} = 0$, $n_0 = 2.44 \times 10^{18}$ m^{-3}, $d = 11.3 \pm 0.5$ μm	[14,16]

An example of typical time dependence of the concentration of mobile ions in liquid crystals doped with contaminated nanoparticles is shown in Figure 1. As can be seen from Figure 1a, the use of contaminated nanoparticles results in the possibility of several regimes, namely the ion-capturing regime (dotted, dashed, and dashed-dotted curves), ion-releasing regime (dashed-dotted-dotted, short-dashed, and short-dotted curves), and no change regime (solid curve). The switching between these regimes is governed by the contamination level of nanoparticles: the ion-capturing regime is observed if $\nu_{NP} < \nu_{NP}^C$, the ion-releasing regime holds true if $\nu_{NP} > \nu_{NP}^C$, and no change regime is reached if $\nu_{NP} = \nu_{NP}^C$, where ν_{NP}^C is the critical contamination factor of nanoparticles defined as $\nu_{NP}^C = \frac{n_0 K_{NP}}{1 + n_0 K_{NP}} \approx n_0 K_{NP}$ where $K_{NP} = \frac{k_a^{NP}}{k_d^{NP}}$. Both ion-releasing and ion-capturing regimes are more pronounced at higher concentration of nanoparticles (Figure 1a). Figure 1a also indicates that the time needed to reach the steady state ($\frac{dn}{dt} = 0$) depends on the concentration of nanoparticles and decreases at higher concentrations.

The kinetics of ion-releasing/ion-capturing processes in liquid crystals doped with nanoparticles is characterized by the time constant τ_{NP} describing how rapidly the steady-state can be reached. This time constant can be defined using a standard definition: $n(\tau_{NP}) - n_0 = (1 - 1/e)(n_\infty - n_0)$, where $n_0 = n(t = 0)$ and $n_\infty = n(t \to \infty)$. According to Equation (3), in the regime of low surface coverage, $\Theta_{NP} << 1$, the time constant can be expressed as $\tau_{NP} = 1/k_d^{NP}(K_{NP}n_{NP}A_{NP}\sigma_S^{NP} + 1)$. Using the relationship between the volume and weight concentration of nanoparticles ($n_{NP} \approx \omega_{NP}\frac{\rho_{LC}}{\rho_{NP}}\frac{1}{V_{NP}}$, where V_{NP} is the volume of a single nanoparticle, and ρ_{LC} (ρ_{NP}) is the density of liquid crystals (nanoparticles)), the time constant can also be rewritten as $\tau_{NP} \approx 1/k_d^{NP}\left(3K_{NP}\sigma_S^{NP}\omega_{NP}\frac{\rho_{LC}}{\rho_{NP}R_{NP}} + 1\right)$, where R_{NP} is the radius of spherical nanoparticles. As can be seen, the time constant τ_{NP} depends on the adsorption-desorption parameters (k_d^{NP}, $K_{NP} = \frac{k_a^{NP}}{k_d^{NP}}$, σ_S^{NP}), the concentration of nanoparticles ω_{NP}, and their size R_{NP}. The dependence of the time constant on the weight concentration of nanoparticles calculated at four different values of their radius is shown in Figure 1b. An increase in the concentration of nanoparticles results in the monotonous decrease of the time constant (Figure 1b). At the same concentration of nanoparticles ω_{NP}, the time constant τ_{NP} is shorter for smaller nanoparticles (Figure 1b).

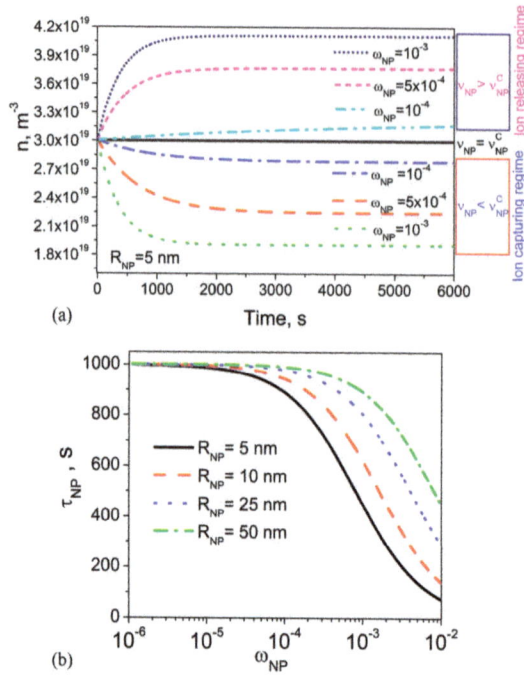

Figure 1. (**a**) The volume concentration of mobile ions n versus time calculated using different values of the weight concentration of nanoparticles ω_{NP} and their contamination factor v_{NP} ($v_{NP} = 10^{-4}$ (dotted, dashed, and dotted-dashed curves); $v_{NP} = 3 \times 10^{-4}$ (solid curve); $v_{NP} = 5 \times 10^{-4}$ (dashed-dotted-dotted, short-dashed, and short-dotted curves)). The radius of nanoparticles R_{NP} is 5 nm; (**b**) The time constant τ_{NP} as a function of the weight concentration of nanoparticles ω_{NP} calculated at different values of the nanoparticle radius R_{NP} ($R_{NP} = 5$ nm (solid curve); $R_{NP} = 10$ nm (dashed curve); $R_{NP} = 25$ nm (dotted curve); $R_{NP} = 50$ nm (dashed-dotted curve)). Other parameters used in simulations: $K_{NP} = 10^{-23}\,\text{m}^3$, $k_d^{NP} = 10^{-3}\,\text{s}^{-1}$, $\sigma_S^{NP} = 0.8 \times 10^{18}\,\text{m}^{-2}$, $n_0 = 3 \times 10^{19}\,\text{m}^{-3}$, $\rho_{NP}/\rho_{LC} = 3.9$.

The time dependence $n(t)$ calculated using the fixed weight concentration of nanoparticles and different values of the nanoparticle radius is shown in Figure 2a. According to this figure, the time needed to reach steady-state depends on the size of nanoparticles. This time (the time constant) is shorter if smaller nanoparticles are used. The dependence of the time constant τ_{NP} on the radius of nanoparticles is shown in Figure 2b. As can be seen, τ_{NP} can be significantly reduced by utilizing smaller nanoparticles and by increasing their concentration.

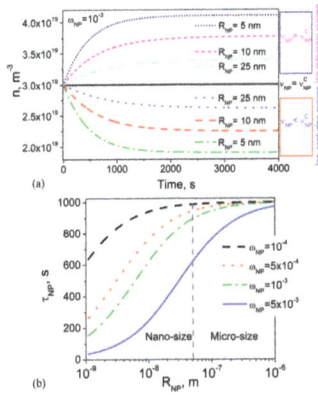

Figure 2. (a) The volume concentration of mobile ions n versus time calculated using different values of the nanoparticle radius R_{NP} ($R_{NP} = 5$ nm (dotted-dashed and short-dotted curves); $R_{NP} = 10$ nm (dashed and short-dashed curves), $R_{NP} = 25$ nm (dotted and dashed-dotted-dotted curves) and their contamination factor v_{NP} ($v_{NP} = 10^{-4}$ (dotted, dashed, and dotted-dashed curves); $v_{NP} = 3 \times 10^{-4}$ (solid curve); $v_{NP} = 5 \times 10^{-4}$ (dashed-dotted-dotted, short-dashed, and short-dotted curves)). The weight concentration of nanoparticles ω_{NP} is 10^{-3}; (b) The time constant τ_{NP} as a function of the radius of nanoparticles R_{NP} calculated at different values of their weight concentration ω_{NP} ($\omega_{NP} = 10^{-4}$ (dashed curve); $\omega_{NP} = 5 \times 10^{-4}$ (dotted curve); $\omega_{NP} = 10^{-3}$ (dashed-dotted curve); $\omega_{NP} = 5 \times 10^{-3}$ (solid curve)). Other parameters used in simulations: $K_{NP} = 10^{-23}$ m^3, $k_d^{NP} = 10^{-3}$ s^{-1}, $\sigma_S^{NP} = 0.8 \times 10^{18}$ m^{-2}, $n_0 = 3 \times 10^{19}$ m^{-3}, $\rho_{NP}/\rho_{LC} = 3.9$.

2.2. The Effects of Contaminated Alignment Layers

The kinetics shown in Figures 1 and 2 was modelled ignoring interactions of ions with alignment layers of the liquid crystal cell. To account for possible effects associated with the adsorption of ions onto the surface of alignment layers, consider sandwich-like cell filled with *pristine* liquid crystals (without nanoparticles). In this case, the change in the concentration of mobile ions in liquid crystals through their adsorption onto the surface of alignment layers can be described by the following rate Equation (4):

$$\frac{dn}{dt} = -k_a^S \frac{\sigma_S^S}{d} n(1 - \Theta_S) + k_d^S \frac{\sigma_S^S}{d} \Theta_S \tag{4}$$

where n is the concentration of mobile ions; σ_S^S is the surface density of all adsorption sites on the surface of alignment layers; d is the cell thickness; k_a^S is the adsorption rate constant describing the ion adsorption onto the surface of alignment layers; and k_d^S is the desorption rate constant; Θ_S is the fractional surface coverage of alignment layers. To compute the time dependence $n(t)$, Equation (4) should be solved together with Equation (5) representing the conservation law of the total number of ions in the liquid crystal cell:

$$n_0 + \frac{\sigma_S^S}{d} v_S = n + \frac{\sigma_S^S}{d} v_S \Theta_S \tag{5}$$

where v_S is the contamination factor of substrates (alignment layers). It equals the fraction of the adsorption sites on the surface of alignment layers occupied by ions-contaminants prior to filling an empty cell with liquid crystals [43,50]. Mathematically, Equations (1), (2), (4) and (5) are similar. In the regime of relatively small surface coverage, $\Theta_S << 1$, an analytical solution can be expressed by Equation (6):

$$n \approx \frac{n_0 + \frac{\sigma_S^S}{d} v_S + \frac{\sigma_S^S K_S}{d}\left(n_0 - \frac{v_S}{K_S}\right) e^{-(k_a^S \frac{\sigma_S^S}{d} + k_d^S)t}}{1 + \frac{\sigma_S^S K_S}{d}} \tag{6}$$

Typical values of physical parameters (σ_S^S, k_d^S, $K_S = \frac{k_a^S}{k_d^S}$) characterizing existing materials are compiled in Table 2.

Table 2. Examples of existing experimental data.

Materials (Liquid Crystals/Films)	Physical Parameters	Ref.
Nematic liquid crystals (ZLI-4792) sandwiched between substrates with alignment layers made of SiO$_x$	$n_0 = 2.4 \times 10^{19}$ m^{-3}, $d = 5 - 6\,\mu$m, $K_s = \frac{k_a}{k_d} \leq 6.7 \times 10^{-22}$ m^3, $\sigma_S \geq 10^{16}$ m^{-2}	[31]
Nematic liquid crystals (ZLI-1132, Merck Corp., Kenilworth, NJ, USA) sandwiched between two substrates with alignment layers (Polyimide AL-1051, JSR Corp, Tokyo, Japan)	$n_0 = 3 \times 10^{19}$ m^{-3}, $d = 4.9\,\mu$m, $k_a n_0 = 4 \times 10^{-4}$ s^{-1}, $k_d = 1.3 \times 10^{-3}$ s^{-1} $\sigma_S = 10^{16} - 10^{18}$ m^{-2}	[35]

The computed time dependence $n(t)$ is shown in Figure 3a.

According to Figure 3a, the type of the observed regime is governed by the ionic contamination of substrates: the ion-capturing regime is reached if $\nu_S < \nu_S^C$ (dotted, dashed, and dashed-dotted curves), the ion-releasing regime takes place if $\nu_S > \nu_S^C$ (dotted-dashed-dashed, short-dashed, and short-dotted curves), and nothing happens if $\nu_S = \nu_S^C$ (solid curve). The time dependence shown in Figure 3a can also be characterized by the time constant τ_S defined as $n(\tau_S) - n_0 = (1 - 1/e)(n_\infty - n_0)$, where $n_0 = n(t = 0)$ and $n_\infty = n(t \to \infty)$. This time constant depends on the cell thickness. It decreases if the cell gap decreases (Figure 3a). According to Equation (6), in the regime of low surface coverage, $\Theta_S << 1$, it can be expressed as $\tau_S = \dfrac{1}{k_d^S \left(\frac{K_S \sigma_S^S}{d} + 1 \right)} = \dfrac{1}{k_d^S \left(\frac{1}{x} + 1 \right)}$. The dependence of

the time constant τ_S on the cell thickness is shown in Figure 3b. As can be seen, the effects of the cell thickness on the time constant are strongly pronounced if relatively thin cells ($d << K_S \sigma_S^S$) are used. These effects become negligible in the case of relatively thick cells ($d >> K_S \sigma_S^S$).

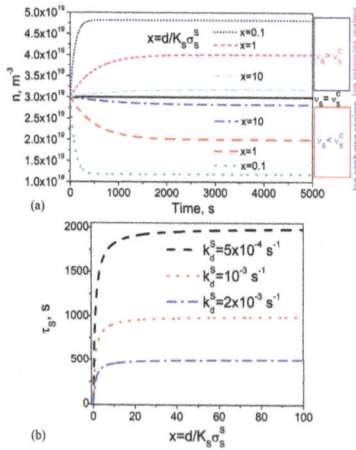

Figure 3. (a) The volume concentration of mobile ions n versus time calculated using different values of the dimensionless parameter x proportional to the cell thickness ($x = \frac{d}{K_S \sigma_S^S}$). The ionic contamination of substrates is quantified by means of the contamination factor ν_S ($\nu_S = 10^{-3}$ (dotted, dashed, and dotted-dashed curves); $\nu_S = 3 \times 10^{-3}$ (solid curve); $\nu_S = 5 \times 10^{-3}$ (dotted-dashed-dashed, short-dashed, and short-dotted curves)). Other parameters used in simulations: $K_S = 10^{-22}$ m^3, $k_d^S = 10^{-3}$ s^{-1}, $\sigma_S^S = 10^{17}$ m^{-2}, $n_0 = 3 \times 10^{19}$ m^{-3}; (b) The time constant τ_S as a function of the dimensionless parameter x calculated at different values of the desorption rate coefficient, k_d^S ($k_d^S = 5 \times 10^{-4}$ s^{-1} (dashed curve); $k_d^S = 10^{-3}$ s^{-1} (dotted curve); $k_d^S = 2 \times 10^{-3}$ s^{-1} (dashed-dotted curve)).

2.3. The Combined Effect of Contaminated Nanoparticles and Substrates

By combining Equations (1) and (4), we can write the generalized rate Equation (7) describing the combined effect of contaminated nanoparticles and alignment layers on the time dependence of the ion-capturing/ion-releasing regimes in liquid crystals:

$$\frac{dn}{dt} = -k_a^{NP} n_{NP} A_{NP} \sigma_S^{NP} n(1 - \Theta_{NP}) + k_d^{NP} n_{NP} A_{NP} \sigma_S^{NP} \Theta_{NP} - k_a^S \frac{\sigma_S^S}{d} n(1 - \Theta_S) + k_d^S \frac{\sigma_S^S}{d} \Theta_S \quad (7)$$

In this case, the kinetics of ion-capturing/ion-releasing processes can be analyzed by solving Equation (7) along with the conservation law of the total number of ions written in more general form (8):

$$n_0 + n_{NP} A_{NP} \sigma_S^{NP} v_{NP} + \frac{\sigma_S^S}{d} v_S = n + n_{NP} A_{NP} \sigma_S^{NP} \Theta_{NP} + \frac{\sigma_S^S}{d} \Theta_S \quad (8)$$

In the regime of relatively low surface coverages, $\Theta_{NP} \ll 1$ and $\Theta_S \ll 1$, an analytical solution can be written as Equation (9):

$$n \approx \frac{n_0 + n_{NP} A_{NP} \sigma_S^{NP} v_{NP} + \frac{\sigma_S^S v_S}{d} + n_{NP} A_{NP} \sigma_S^{NP} K_{NP} \left(n_0 - \frac{v_{NP}}{K_{NP}}\right) e^{-(k_a^{NP} n_{NP} A_{NP} \sigma_S^{NP} + k_d^{NP})t} + \frac{\sigma_S^S K_S}{d}\left(n_0 - \frac{v_S}{K_S}\right) e^{-(k_a^S \frac{\sigma_S^S}{d} + k_d^S)t}}{1 + K_{NP} n_{NP} A_{NP} \sigma_S^{NP} + \frac{\sigma_S^S K_S}{d}} \quad (9)$$

An example of the time dependence of ion-capturing/ion-releasing regimes in liquid crystals doped with contaminated nanoparticles and sandwiched between contaminated substrates is shown in Figure 4.

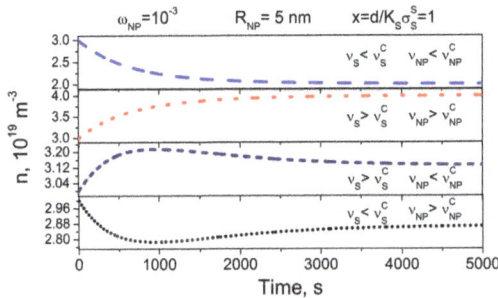

Figure 4. Time dependence of the volume concentration of mobile ions n in liquid crystals doped with contaminated nanoparticles and sandwiched between contaminated substrates. The radius of nanoparticles is 5 nm, and their weight concentration is 0.001. The dimensionless parameter $x = 1$. The contamination factors of both nanoparticles and substrates are varied: dashed curve ($v_S = 10^{-3} < v_S^C = 3 \times 10^{-3}$, $v_{NP} = 2 \times 10^{-4} < v_{NP}^C = 3 \times 10^{-4}$); dotted curve ($v_S = 5 \times 10^{-3} > v_S^C = 3 \times 10^{-3}$, $v_{NP} = 4 \times 10^{-4} > v_{NP}^C = 3 \times 10^{-4}$); short-dashed ($v_S = 5 \times 10^{-3} > v_S^C = 3 \times 10^{-3}$, $v_{NP} = 2 \times 10^{-4} < v_{NP}^C = 3 \times 10^{-4}$); short-dotted ($v_S = 10^{-3} < v_S^C = 3 \times 10^{-3}$, $v_{NP} = 4 \times 10^{-4} > v_{NP}^C = 3 \times 10^{-4}$). Other parameters used in simulations: $K_S = 10^{-22}$ m^3, $K_{NP} = 10^{-23}$ m^3, $k_d^S = k_d^{NP} = 10^{-3}$ s^{-1}, $\sigma_S^S = 10^{17}$ m^{-2}, $\sigma_S^{NP} = 10^{18}$ m^{-2}, $\rho_{NP}/\rho_{LC} = 3.9$, $n_0 = 3 \times 10^{19}$ m^{-3}.

The type of the observed regime depends on the interplay between the contamination factor of nanoparticles (v_{NP}) and substrates (v_S), adsorption/desorption parameters (k_a^{NP}, k_d^{NP}, $K_{NP} = k_a^{NP}/k_d^{NP}$, k_a^S, k_d^S, $K_S = k_a^S/k_d^S$, σ_S^{NP}, σ_S^S), the concentration of nanoparticles (ω_{NP}) and their size (R_{NP}), initial concentration of ions in liquid crystals (n_0), and cell thickness (d).

The time dependence of the ion-capturing regime is represented by dashed and short-dotted curves in Figure 4. Dotted and short-dashed curves show the kinetics of the ion-releasing regime (Figure 4). The kinetics of ion-capturing/ion-releasing processes is characterized by two time constants,

$$\tau_{NP} = 1/k_d^{NP}\left(K_{NP}n_{NP}A_{NP}\sigma_S^{NP}+1\right) \text{ and } \tau_S = \frac{1}{k_d^S\left(\frac{K_S\sigma_S^S}{d}+1\right)},$$ originated from the presence of nanoparticles and substrates, respectively. An interesting feature is the possibility of both monotonous (Figure 4, dashed and dotted curves) and non-monotonous (Figure 4, short-dotted and short-dashed curves) time dependence $n(t)$.

3. Conclusions

Since the liquid crystal cell is a major component of practically any electro-optical device utilizing mesogenic materials, the results presented in this paper have important practical implications. Once the cell is filled with liquid crystals, interactions of ions with alignment films and/or nanodopants through the ion adsorption/ion desorption processes result in a strongly pronounced time dependence of its electrical properties. Electro-optical response of liquid crystals, especially at relatively low frequencies, can be affected by the presence of ions in liquid crystals [2,3,51]. As a result, electro-optical properties of the liquid crystal device can also become time-dependent. The knowledge of this time dependence is very important from both scientific and applied perspectives. Findings presented in this paper provide an important information on the kinetics of ion-capturing/ion-releasing processes in liquid crystal cells utilizing nanomaterials and alignment layers. Very important aspect of the present study is the consideration of the ionic contamination of both nanodopants and alignment films. Too often this factor (ionic contamination) is overlooked in existing literature. The presented model shows possible scenarios of time-dependent ion-capturing/ion-releasing regimes in liquid crystal devices and their dependence on the ionic contamination of both nanodopants and alignment layers. Thus, it can be used for the analysis of existing experimental data and will also guide the design of liquid crystal devices utilizing nanoparticles and alignment films.

Some limitations of the proposed model should also be mentioned. The limits and applicability of this approach have already been discussed [41–43]. In addition, this model does not consider the nature and origin of ionic contaminants. Different types of materials (liquid crystals, nanodopants, alignment films) and ionic contaminants can be distinguished by using different values of physical parameters used in the model (k_d^{NP}, $K_{NP} = \frac{k_a^{NP}}{k_d^{NP}}$, σ_S^{NP}, k_d^S, $K_S = \frac{k_a^S}{k_d^S}$, σ_S^S). The geometry of the cell and orientation of liquid crystals is fixed thus the afore-mentioned physical parameters characterizing the ion-capturing/ion-releasing processes are considered constant.

To summarize, the kinetics of ion-capturing/ion-releasing processes in liquid crystal cells utilizing contaminated nanoparticles and alignment layers exhibits several non-trivial features (Figures 1–4). These features originate from the presence of both nanoparticles and alignment layers and from their ionic contamination. The ionic contamination of nanoparticles and alignment layers governs the type of the regime (Figures 1–3) and determines whether the observed time dependence $n(t)$ is monotonous or non-monotonous (Figure 4). This time dependence $n(t)$ is described by time constants τ_{NP} and τ_S characterizing ion adsorption/ion desorption processes on the surface of nanoparticles and alignment layers, respectively. The time constant τ_{NP} depends on the ion-nanoparticle related adsorption parameters (k_d^{NP}, $K_{NP} = \frac{k_a^{NP}}{k_d^{NP}}$, σ_S^{NP}), concentration of nanoparticles ω_{NP}, and their size R_{NP} (Figures 1 and 2). The time constant τ_S is also a function of the adsorption parameters characterizing the ion adsorption/ion desorption processes on the surface of alignment layers (k_d^S, $K_S = \frac{k_a^S}{k_d^S}$, σ_S^S). In addition, it depends on the thickness of the cell (Figure 3). The obtained results offer an efficient way to control the kinetics of ion-capturing/ion-releasing process in liquid crystal devices by changing the concentration of nanoparticles, their size, cell thickness, and the ionic contamination of both nanoparticles and alignment layers.

Acknowledgments: The author would like to acknowledge the support provided by the UCCS BioFrontiers Center at the University of Colorado Colorado Springs.

Conflicts of Interest: The author declares no conflict of interest.

References

1. Yang, D.-K.; Wu, S.-T. *Liquid Crystal Devices*; John Wiley & Sons: Hoboken, NJ, USA, 2006; pp. 1–378.
2. Chigrinov, V.G. *Liquid Crystal Devices: Physics and Applications*; Artech House: Boston, MA, USA, 1999; pp. 1–360.
3. Naemura, S. Electrical properties of liquid crystal materials for display applications. *Mater. Res. Soc. Symp. Proc.* **1999**, *559*, 263–274. [CrossRef]
4. Geis, M.W.; Bos, P.J.; Liberman, V.; Rothschild, M. Broadband optical switch based on liquid crystal dynamic scattering. *Opt. Express* **2016**, *24*, 13812–13823. [CrossRef] [PubMed]
5. Serak, S.V.; Hrozhyk, U.; Hwang, J.; Tabiryan, N.V.; Steeves, D.; Kimball, B.R. High contrast switching of transmission due to electrohydrodynamic effect in stacked thin systems of liquid crystals. *Appl. Opt.* **2016**, *55*, 8506–8512. [CrossRef] [PubMed]
6. Konshina, E.A.; Shcherbinin, D.P. Study of dynamic light scattering in nematic liquid crystal and its optical, electrical and switching characteristics. *Liq. Cryst.* **2017**. [CrossRef]
7. Garbovskiy, Y.; Glushchenko, I. Nano-objects and ions in liquid crystals: Ion trapping effect and related phenomena. *Crystals* **2015**, *5*, 501–533. [CrossRef]
8. Wu, P.C.; Lisetski, L.N.; Lee, W. Suppressed ionic effect and low-frequency texture transitions in a cholesteric liquid crystal doped with graphene nanoplatelets. *Opt. Express* **2015**, *23*, 11195–11204. [CrossRef] [PubMed]
9. Mun, H.-Y.; Park, H.-G.; Jeong, H.-C.; Lee, J.H.; Oh, B.Y.; Seo, D.-S. Thermal and electro-optical properties of cerium-oxide-doped liquid-crystal devices. *Liq. Cryst.* **2017**, *44*, 538–543. [CrossRef]
10. Shcherbinin, D.P.; Konshina, E.A. Ionic impurities in nematic liquid crystal doped with quantum dots CdSe/ZnS. *Liq. Cryst.* **2017**, *44*, 648–655. [CrossRef]
11. Sharma, K.P.; Malik, P.; Raina, K.K. Electro-optic, dielectric and optical studies of $NiFe_2O_4$-ferroelectric liquid crystal: A soft magnetoelectric material. *Liq. Cryst.* **2016**, *43*, 1671–1681.
12. Podgornov, F.V.; Wipf, R.; Stühn, B.; Ryzhkova, A.V.; Haase, W. Low-frequency relaxation modes in ferroelectric liquid crystal/gold nanoparticle dispersion: Impact of nanoparticle shape. *Liq. Cryst.* **2016**, *43*, 1536–1547. [CrossRef]
13. Garbovskiy, Y.; Glushchenko, I. Ion trapping by means of ferroelectric nanoparticles, and the quantification of this process in liquid crystals. *Appl. Phys. Lett.* **2015**, *107*, 041106. [CrossRef]
14. Hsiao, Y.G.; Huang, S.M.; Yeh, E.R.; Lee, W. Temperature-dependent electrical and dielectric properties of nematic liquid crystals doped with ferroelectric particles. *Displays* **2016**, *44*, 61–65. [CrossRef]
15. Garbovskiy, Y. Switching between purification and contamination regimes governed by the ionic purity of nanoparticles dispersed in liquid crystals. *Appl. Phys. Lett.* **2016**, *108*, 121104. [CrossRef]
16. Garbovskiy, Y. Electrical properties of liquid crystal nano-colloids analysed from perspectives of the ionic purity of nano-dopants. *Liq. Cryst.* **2016**, *43*, 648–653. [CrossRef]
17. Tomylko, S.; Yaroshchuk, O.; Kovalchuk, O.; Maschke, U.; Yamaguchi, R. Dielectric properties of nematic liquid crystal modified with diamond nanoparticles. *Ukrainian J. Phys.* **2012**, *57*, 239–243.
18. Samoilov, A.N.; Minenko, S.S.; Fedoryako, A.P.; Lisetski, L.N.; Lebovka, N.I.; Soskin, M.S. Multi-walled vs. single-walled carbon nanotube dispersions in nematic liquid crystals: Comparative studies of optical transmission and dielectric properties. *Funct. Mater.* **2014**, *21*, 190–194. [CrossRef]
19. Yadav, S.P.; Manohar, R.; Singh, S. Effect of TiO_2 nanoparticles dispersion on ionic behaviour in nematic liquid crystal. *Liq. Cryst.* **2015**, *42*, 1095–1101. [CrossRef]
20. Garbovskiy, Y. Impact of contaminated nanoparticles on the non-monotonous change in the concentration of mobile ions in liquid crystals. *Liq. Cryst.* **2016**, *43*, 664–670. [CrossRef]
21. Garbovskiy, Y. Adsorption of ions onto nanosolids dispersed in liquid crystals: Towards understanding the ion trapping effect in nanocolloids. *Chem. Phys. Lett.* **2016**, *651*, 144–147. [CrossRef]
22. Urbanski, M.; Lagerwall, J.P.F. Why organically functionalized nanoparticles increase the electrical conductivity of nematic liquid crystal dispersions. *J. Mater. Chem. C* **2017**, *5*, 8802–8809. [CrossRef]
23. Garbovskiy, Y. Nanoparticle enabled thermal control of ions in liquid crystals. *Liq. Cryst.* **2017**, *44*, 948–955. [CrossRef]
24. Garbovskiy, Y. Ions in liquid crystals doped with nanoparticles: Conventional and counterintuitive temperature effects. *Liq. Cryst.* **2017**, *44*, 1402–1408. [CrossRef]

25. Liu, H.; Lee, W. Time-varying ionic properties of a liquid-crystal cell. *Appl. Phys. Lett.* **2010**, *97*, 023510. [CrossRef]

26. Wu, P.-C.; Yang, S.-Y.; Lee, W. Recovery of UV-degraded electrical properties of nematic liquid crystals doped with TiO_2 nanoparticles. *J. Mol. Liq.* **2016**, *218*, 150–155. [CrossRef]

27. Takatoh, K.; Hasegawa, M.; Koden, M.; Iton, N.; Hasegawa, R.; Sakamoto, M. *Alignment Technologies and Applications of Liquid Crystal Devices*; Taylor & Francis: New York, NY, USA, 2005; pp. 1–320.

28. Furuichi, K.; Xu, J.; Furuta, H.; Kobayashi, S.; Yoshida, N.; Tounai, A.; Tanaka, Y. 38.4: Effect of Ion Capturing Films on the EO Characteristics of Polymer-Stabilized V-FLCD. *SID Symp. Dig. Tech. Pap.* **2002**, *33*, 1114–1117. [CrossRef]

29. Furuichi, K.; Xu, J.; Inoue, M.; Furuta, H.; Yoshida, N.; Tounai, A.; Tanaka, Y.; Mochizuki, A.; Kobayashi, S. Effect of Ion Trapping Films on the Electrooptic Characteristics of Polymer-Stabilized Ferroelectric Liquid Crystal Display Exhibiting V-Shaped Switching. *Jpn. J. Appl. Phys.* **2003**, *42*, 4411–4415. [CrossRef]

30. Kobayashi, S.; Xu, J.; Furuta, H.; Murakami, Y.; Kawamoto, S.; Ohkouchi, M.; Hasebe, H.; Takatsu, H. Fabrication and electro-optic characteristics of polymer-stabilized V-mode ferroelectric liquid crystal display and intrinsic H-V-mode ferroelectric liquid crystal displays: Their application to field sequential full colour active matrix liquid crystal displays. *Opt. Eng.* **2004**, *43*, 290–298.

31. Huang, Y.; Bos, P.J.; Bhowmik, A. The ion capturing effect of 5 SiO_x alignment films in liquid crystal devices. *J. Appl. Phys.* **2010**, *108*, 064502. [CrossRef]

32. Huang, Y.; Bhowmik, A.; Bos, P.J. Characterization of Ionic Impurities Adsorbed onto a 5° SiO_x Alignment Film. *Jpn. J. Appl. Phys.* **2012**, *51*, 031701.

33. Huang, Y.; Bhowmik, A.; Bos, P.J. The effect of salt on ion adsorption on a SiO_x alignment film and reduced conductivity of a liquid crystal host. *J. Appl. Phys.* **2012**, *111*, 024501. [CrossRef]

34. Basu, R.; Lee, A. Ion trapping by the graphene electrode in a graphene-ITO hybrid liquid crystal cell. *Appl. Phys. Lett.* **2017**, *111*, 161905. [CrossRef]

35. Mizusaki, M.; Miyashita, T.; Uchida, T.; Yamada, Y.; Ishii, Y.; Mizushima, S. Generation mechanism of residual direct current voltage in a liquid crystal display and its evaluation parameters related to liquid crystal and alignment layer materials. *J. Appl. Phys.* **2007**, *102*, 014904. [CrossRef]

36. Mizusaki, M.; Miyashita, T.; Uchida, T. Behavior of ion affecting image sticking on liquid crystal displays under application of direct current voltage. *J. Appl. Phys.* **2010**, *108*, 104903. [CrossRef]

37. Mizusaki, M.; Miyashita, T.; Uchida, T. Kinetic analysis of image sticking with adsorption and desorption of ions to a surface of an alignment layer. *J. Appl. Phys.* **2012**, *112*, 044510. [CrossRef]

38. Mizusaki, M.; Yoshimura, Y.; Yamada, Y.; Okamoto, K. Analysis of ion behavior affecting voltage holding property of liquid crystal displays. *Jpn. J. Appl. Phys.* **2012**, *51*, 014102. [CrossRef]

39. Xu, D.; Peng, F.; Chen, H.; Yuan, J.; Wu, S.-T.; Li, M.-C.; Lee, S.-L.; Tsai, W.-C. Image sticking in liquid crystal displays with lateral electric fields. *J. Appl. Phys.* **2014**, *116*, 193102. [CrossRef]

40. Barbero, G.; Evangelista, L.R. *Adsorption Phenomena and Anchoring Energy in Nematic Liquid Crystals*; Taylor & Francis: Boca Raton, FL, USA, 2006; pp. 1–352.

41. Garbovskiy, Y. Adsorption/desorption of ions in liquid crystal nano-colloids: The applicability of the Langmuir isotherm, impact of high electric fields, and effects of the nanoparticle's size. *Liq. Cryst.* **2016**, *43*, 853–860. [CrossRef]

42. Garbovskiy, Y. The purification and contamination of liquid crystals by means of nanoparticles. The case of weakly ionized species. *Chem. Phys. Lett.* **2016**, *658*, 331–335. [CrossRef]

43. Garbovskiy, Y. Ions and size effects in nanoparticle/liquid crystal colloids sandwiched between two substrates. The case of two types of fully ionized species. *Chem. Phys. Lett.* **2017**, *679*, 77–85. [CrossRef]

44. Riley, K.F.; Hobson, M.P.; Bence, S.J. *Mathematical Methods for Physics and Engineering*; Cambridge University Press: New York, NY, USA, 1997; pp. 1–1008.

45. Marczewski, A.W. Analysis of kinetic Langmuir model. Part I: Integrated kinetic Langmuir equation (IKL): A new complete analytical solution of the Langmuir rate equation. *Langmuir* **2010**, *26*, 15229–15238. [CrossRef] [PubMed]

46. Gonen, Y.; Rytwo, G. A full analytical solution for the sorption–desorption kinetic process related to Langmuir equilibrium conditions. *J. Phys. Chem. C* **2007**, *111*, 1816–1819. [CrossRef]

47. Tang, C.Y.; Huang, S.M.; Lee, W. Electrical properties of nematic liquid crystals doped with anatase TiO_2 nanoparticles. *J. Phys. D Appl. Phys.* **2011**, *44*, 355102. [CrossRef]

48. Jian, B.R.; Tang, C.Y.; Lee, W. Temperature-dependent electrical properties of dilute suspensions of carbon nanotubes in nematic liquid crystals. *Carbon* **2011**, *49*, 910–914. [CrossRef]

49. Wu, P.W.; Lee, W. Phase and dielectric behaviors of a polymorphic liquid crystal doped with graphene nanoplatelets. *Appl. Phys. Lett.* **2013**, *102*, 162904. [CrossRef]

50. Garbovskiy, Y. Ion capturing/ion releasing films and nanoparticles in liquid crystal devices. *Appl. Phys. Lett.* **2017**, *110*, 041103. [CrossRef]

51. Ciuchi, F.; Mazzulla, A.; Pane, A.; Adrian Reyes, J. ac and dc electro-optical response of planar aligned liquid crystal cells. *Appl. Phys. Lett.* **2007**, *91*, 232902. [CrossRef]

nanomaterials

MDPI

Article

Dynamic Response of Graphitic Flakes in Nematic Liquid Crystals: Confinement and Host Effect

Weiwei Tie [1,2], Surjya Sarathi Bhattacharyya [1,3], Yuanhao Gao [1], Zhi Zheng [1], Eun Jeong Shin [2], Tae Hyung Kim [2], MinSu Kim [2], Joong Hee Lee [2] and Seung Hee Lee [2,*]

[1] Key Laboratory of Micro-Nano Materials for Energy Storage and Conversion of Henan Province, Institute of Surface Micro and Nano Materials, College of Advanced Materials and Energy, Xuchang University, Xuchang 461000, Henan, China; dwtie929@hotmail.com (W.T.); surjyasarathi@gmail.com (S.S.B.); gyh-2007@sohu.com (Y.G.); zhengzhi99999@gmail.com (Z.Z.)

[2] Applied Materials Institute for BIN Convergence, Department of BIN Convergence Technology and Graduate School of Printable and Flexible Electronics, Chonbuk National University, Jeonju 561-756, Jeonbuk, Korea; zxcvb823@naver.com (E.J.S.); kth1811@naver.com (T.H.K.); mkim12@kent.edu (M.K.); jhl@jbnu.ac.kr (J.H.L.)

[3] Asutosh College, 92, Shyamaprasad Mukherjee Road, Kolkata 700 026, West Bengal, India

[*] Correspondence: lsh1@chonbuk.ac.kr; Tel.: +82-63-270-2443

Received: 22 July 2017; Accepted: 30 August 2017; Published: 1 September 2017

Abstract: Electric field-induced reorientation of suspended graphitic (GP) flakes and its relaxation back to the original state in a nematic liquid crystal (NLC) host are of interest not only in academia, but also in industrial applications, such as polarizer-free and optical film-free displays, and electro-optic light modulators. As the phenomenon has been demonstrated by thorough observation, the detailed study of the physical properties of the host NLC (the magnitude of dielectric anisotropy, elastic constants, and rotational viscosity), the size of the GP flakes, and cell thickness, are urgently required to be explored and investigated. Here, we demonstrate that the response time of GP flakes reorientation associated with an NLC host can be effectively enhanced by controlling the physical properties. In a vertical field-on state, higher dielectric anisotropy and higher elasticity of NLC give rise to quicker reorientation of the GP flakes (switching from planar to vertical alignment) due to the field-induced coupling effect of interfacial Maxwell-Wagner polarization and NLC reorientation. In a field off-state, lower rotational viscosity of NLC and lower cell thickness can help to reduce the decay time of GP flakes reoriented from vertical to planar alignment. This is mainly attributed to strong coupling between GP flakes and NLC originating from the strong π-π interaction between benzene rings in the honeycomb-like graphene structure and in NLC molecules. The high-uniformity of reoriented GP flakes exhibits a possibility of new light modulation with a relatively faster response time in the switching process and, thus, it can show potential application in field-induced memory and modulation devices.

Keywords: graphitic flakes; liquid crystal; Maxwell-Wagner polarization; dynamic response

1. Introduction

The importance of electric field-induced reorientation of non-spherical particles has gradually emerged, such that polymer cholesteric liquid crystal flakes (PCLC) for electro-optical devices and displays have been widely studied [1,2]. The PCLC flakes in a fluid opened a possibility to realize a particle-based and electric field-driven technology into a wide variety of applications such as displays and electronic papers with no optical polarizers and filters. Despite of the immense possibility of various applications, the development in switching technology of PCLC flakes is still in an early stage. For instance, the uniformity of reorientation needs to be improved and the relaxation time is required to be shortened.

Recently, an optically-opaque two-dimensional (2D) graphitic (GP) flake, which consists of multi-layered graphene sheets by π-π stacking, can provide an opportunity for a new light modulation technology [3–5]. Such technology can give rise to sufficient darkness in a field-off state due to the planar configuration of GP flakes in nematic liquid crystals (NLCs) with respect to the substrate. Under an applied electric field as low as tens of millivolts per micrometer, GP flakes reorient almost 90° from a planar face-up configuration to a perpendicular edge-up configuration, and a transmitting bright state is realized [5]. This switching technology upon electric field-induced reorientation of GP flakes relies on a combined interaction of the NLC director reorientation and Maxwell-Wagner polarization. Recent investigation over graphene oxide/thermotropic NLC-dispersed systems [6] has revealed electro-optical properties, viz. the threshold voltage and splay elastic constant of such systems strongly increase with increasing graphene oxide concentration. However, we have realized much reliable and reproducible switching technology for practical application devoid of electric field-induced flake aggregation. Information displays or electronic papers assisted by GP flake switching could realize low power consumption, high brightness, and multi-color capability without the use of filters and polarizers, which have great potential for both commercial and scientific applications.

The response time for reorientation of NLC can be described as:

$$\tau_{decay} = \frac{\gamma_1 d^2}{\pi^2 K} = \frac{\gamma_1}{\varepsilon_0 |\Delta \varepsilon| E_c^2} \tag{1}$$

and:

$$\tau_{rise} = \frac{\gamma_1}{\varepsilon_0 |\Delta \varepsilon| E^2 - (\pi^2/d^2)K} = \frac{\gamma_1}{\varepsilon_0 |\Delta \varepsilon| (E^2 - E_c^2)} \tag{2}$$

where γ_1, d, K, ε_0, and $\Delta \varepsilon$ denote physical properties of a liquid crystal, viz. the rotational viscosity, the cell thickness, the elastic constant, the dielectric constant in vacuum, and the dielectric anisotropy, respectively, and E_c, and E denote the critical electric field and an external electric field, respectively [6–10]. This implies that reducing γ_1 and d, and increasing K and $\Delta \varepsilon$ will contribute to a decrease in the response time of liquid crystals. GP flakes, while suspended in liquid crystal materials, reorient and relax following the liquid crystal director [5,10]. The simplified model to describe such an inhomogeneous configuration can be a double-layered structure with different dielectric permittivity ε and conductivity σ of each material, and the relaxation frequency can be described as [10]:

$$\tau = \kappa \varepsilon_0 \frac{\varepsilon_1 + \varepsilon_2}{\sigma_1 + \sigma_2} \tag{3}$$

This interaction can be further explained as the electric field-induced potential energy of GP flakes coupling with the contribution of positive NLC reorientation to generate the rotational kinetic energy for flake reorientation [5]. More importantly, the field-induced reorientation of GP flakes relaxes back to their initial planar face-up state for recovering the dark field-off state aided by the NLC director reorientation. This reversible reorientation of GP flakes allows optical modulation controlled by an electric field signal. Although the total response time, i.e., the time to complete both the reorientation and relaxation process is faster than PCLC flakes in general [1,2], the response time of the GP flakes still needs to be further improved for practical applications. We have carefully studied the dynamic reorientation phenomenon of graphene/graphitic flakes in NLCs under an electric field [5]. In the field-on state, we have found that the reorientation time of the GP device can be efficiently improved by optimizing several parameters, such as the electric field strength, the flake anisotropy, as well as the flake size and, again, the reorientation depends on the field-induced synergistic effect of interfacial Maxwell-Wagner polarization and NLC director reorientation. It is also found that π-π interaction between the GP surface and NLC molecule is so strong that the surface anchoring between the NLC molecules and GP flakes will allow such GP flakes following the NLC directors to reorient it collaboratively [7,8,11–13]. Thus, further investigation on the contribution of NLC director reorientation to the GP flake reorientation and the relaxation process is urgently required.

The response time of such system is defined in [5], and it can be further modified as:

$$\tau \approx \frac{2\pi\sqrt{\rho a}R_g}{\sqrt{\left(\frac{4\pi}{3}l_iB_iE^2 + 2AK\right) - \frac{\gamma_1^2}{4\rho a R_g^2}}},$$

(4)

where notations represent that the mass per unit area of flakes ρ, the projected area of flakes a, radius of gyration R_g, dimension of flakes in the axes ($i = x, y, z$) l, and a multiplicative constant A. The Clausius-Mosotti factor is defined by:

$$B_i = \frac{\left(\varepsilon_p - \varepsilon_h\right)}{\left[\varepsilon_h + L_i\left(\varepsilon_p - \varepsilon_h\right)\right]},$$

where L_i is the geometrical depolarization factor. Therefore, variables in Equation (4), such as the physical properties of the NLC host, size, and aspect ratio of GP flakes, will provide a detailed contribution to the response time of the reorientation, as well as the relaxation process of the GP flakes.

2. Materials and Methods

The functionalized GP flakes have been synthesized through room-temperature oxidation of precursor graphite to generate graphite oxide and subsequent low-temperature thermal exfoliation to obtain functionalized graphene. The synthesized functionalized graphene mostly consists of an exfoliated graphene layer and some portion of the stacked layer structures with an expanded interlayer spacing of 7.5 Å. The named functionalized graphitic flake generated in the present synthesis scheme contains fewer defects over graphene sheets (relatively unbroken sp^2 hybridization carbon atom over graphene plane) and less oxygen content over the graphene plane in comparison with commercial graphene oxide. The shape, uniformity, and internal structure of the flakes have been characterized using optical microscopy (OM) and scanning electron microscopy (SEM). The synthesis and detailed physical characterization of the GP flakes used in present investigation have been described elsewhere [3,14].

Test cells named as Case I, Case II and Case III shown in Table 1 were prepared in order to investigate the NLC host type and cell thickness-dependent reorientation and relaxation behavior of the GP flakes in NLC by comparing with the previous results as in [5]. Indium-tin-oxide (ITO) is used as electrodes on transparent glass substrates. Homogeneous alignment layers on both substrates are made by spin-coating a polyimide solution, viz. SE-6514 (Japan Synthetic Rubber Co., Tokyo, Japan), and then the substrates are baked at 70 °C for 5 min and 200 °C for one hour. The baked alignment layers are uniformly rubbed by a rubbing machine. The GP flakes and NLC suspension is sandwiched by the rubbed ITO substrates in anti-parallel direction and the uniform cell thickness is maintained at ~44 μm using a film spacer between the substrates. The cell thickness has been measured by comparing the capacitance of an empty cell and a benzene-filled cell using a 4284A Inductance capacitance resistance (LCR-meter) (Agilent Technologies, Santa Clara, CA, USA) [3,5,15].

Two commercially-available positive dielectric anisotropic NLC mixtures with different physical properties (dielectric anisotropy $\Delta\varepsilon = 5.3$, rotational viscosity $\gamma = 132$ mPa·s, splay elastic constant $K_{11} = 13.2 \times 10^{-12}$ N), viz. ZLI-4792, and ($\Delta\varepsilon = 14.8$, $\gamma = 148$ mPa·s, $K_{11} = 9.3 \times 10^{-12}$ N), viz. ZLI-4535 (Merck, Tokyo, Japan), have been used for present investigation. The GP flake-NLC suspensions have been prepared by adding ~0.005 wt % of flake powders directly into the NLC mixtures. In order to achieve the homogeneity of the dispersion of flakes with minimum damage, the suspensions are ultrasonicated for one hour, keeping the temperature at 30 °C (frequency, 28 kHz; input power, 60 W).

Table 1. Prepared test cells.

Samples	LC	Cell Thickness (μm)
Case I	ZLI-4792	44
Case II	ZLI-4535	44
Case III (from reference [5])	ZLI-4792	70

The used GP flake-NLC suspensions with ZLI-4792 and ZLI-4535 are filled into test cells and named as Case I, Case II, and Case III [5], respectively, as shown in Table 2. Samples are filled into the mentioned cells by capillary action at room temperature. Although most of the GP flakes in the GP-NLC suspension are supposed to be single-layered, or composed of a few layers of graphene after exfoliation, some thicker graphitic stacks do remain suspended in the mixture, exhibiting the desired dark state. They are characterized by optical microscopy, having significantly distinct thickness formed by several hundred layers forming an optically opaque state, and suspended in the planar face-up configuration in NLCs with an area of 10–300 μm^2 and, hence, a volume of 10–450 μm^3. The sample textures are captured by a Panasonic 5100 digital camera with a time resolution of 100 ms attached with optical microscope (OM, Nikon DXM1200, Tokyo, Japan) while test cells are driven by an AC field (0 to 200 mV/μm) with a constant driving frequency (60 Hz) using an arbitrary function generator (Tektronix AFG3022, Beaverton, OR, USA). The transmitted light intensity of the captured images has been analyzed with the help of IMT *i*-Solution software (Image and Microscope Technology Co.).

Table 2. Measured GP flakes of Case I, Case II and Case III.

Samples	Flake Area (μm^2)	Flake Aspect Ratio (μm)
Case I	Case 1: 35 Case 2: 85	Case 1: 2:1 Case 2: 1:1
Case II	Case 1: 37 Case 2: 90	Case 2: 2:1 Case 2: 1:1
Case III (from reference [5])	Case 1: 65 Case 2: 85	Case 2: 2:1 Case 2: 2.6:1

3. Results and Discussion

The micrometer-sized GP flakes dispersed in the homogeneously-aligned NLC host in the mentioned cell geometry have been observed under an optical microscope. Figure 1 shows the representative case study of arbitrarily-shaped GP flakes' reorientation for comparable morphology at field-off and field-on states with a fixed AC electric field strength of 90.0 mV/μm. Areas of suspended GP flakes in the field-off planar face-up configuration are microscopically measured to be 35 and 85 μm^2 in the NLC host of ZIL4792 (a,b) (Case I) and those for the GP flake counterpart are 37 and 90 μm^2 in the NLC host of ZIL4535 (c,d) (Case II), respectively. Figure 1 clearly displays all GP flakes reorient from face-up to almost edge-up configuration around their long axis parallel to the applied field direction at field-on state. Once the driving field is removed, the GP flakes relax back to their planar initial state. This observed reorientation behavior is reliable and reproducible under hundreds of trials.

Figure 2 depicts the field-dependent variation of transmitted light intensity surrounding GP flakes. The optical transmittance has been measured locally on regions surrounding the GP flake in an area ~20 μm × 20 μm from optical microscopic images using *i*-Solution image analysis software and data points are plotted while the solid line represents a guide for the eye to the experimental data. Initially, transmission intensity is low as most of the transmission is blocked by the considerably large, effectively exposed area of the opaque flake in the direction of propagation of light. As the flakes reorient with the increasing applied field, the transmission gradually increases and, finally, reach their maximum transmittance for each case. Normally, a different maximum transmission intensity is ascribed to the different thicknesses of opaque graphitic flakes. In consideration of the field strength dependent reorientation time previously discussed in detail [5], we have investigated the influence of cell confinement on the response time of GP flakes in the next section.

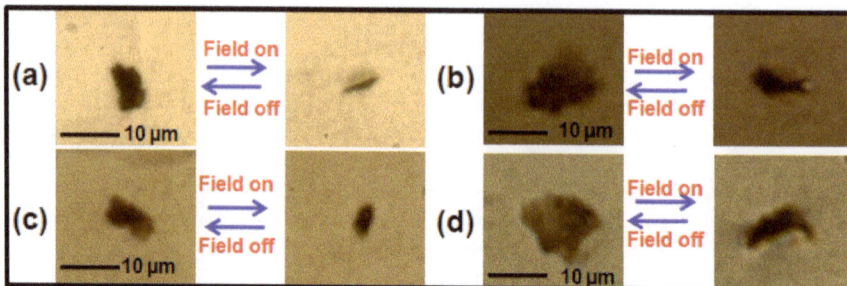

Figure 1. (Color online) Optical microphotographs exhibiting field-induced reorientation of representative GP flakes with similar flake areas and aspect ratios in low (**a**,**b**) and high (**c**,**d**) dielectric anisotropy nematic liquid crystal. (**a–d**) Case I: flake area and aspect ratio in (**a**,**b**): 35 μm^2 with an aspect ratio of 2:1 and 85 μm^2 with an aspect ratio of 1:1; Case II: flake area and aspect ratio in (**c**,**d**): 37 μm^2 with an aspect ratio of 2:1 and 90 μm^2 with an aspect ratio of 1:1.

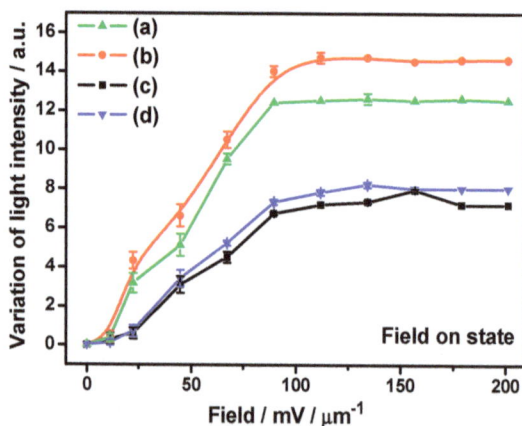

Figure 2. The measured reorientation-induced light intensity variation of representative GP flakes according to Figure 1. The applied electric fields for Case I and Case II are 0, 22, 44, 67, 90, 100, 112, 134, 156, 180, and 200 mV/μm, respectively (Case I: flake area and aspect ratio in (**a**,**b**): 35 μm^2 with an aspect ratio of 2:1 and 85 μm^2 with an aspect ratio of 1:1; Case II: flake area and aspect ratio in (**c**,**d**): 37 μm^2 with an aspect ratio of 2:1 and 90 μm^2 with an aspect ratio of 1:1). The scattered points represent experimental data along with measurement error bars and the lines are guides for the eye. Error bars for respective data points are assigned considering standard deviations obtained in the repetitive observations.

Figure 3 shows variation of transmitted light intensity as a function of relaxation time of functionalized graphene flakes in the LC matrix. It is interesting to mention here that the transmission intensity around GP flakes is found to be independent of the electric field. The flakes gradually recover to their initial state and, hence, the corresponding transmittance decreases after field removal. Experimental data is represented as scattered points with the solid line as a guide for the eye. However, it is observed that the flakes suspended in different NLC hosts recover to their initial state with different relaxation times, which will be mainly discussed in the following section. The mentioned observation confirms the reversible reorientation of GP flakes according to the optical images of Figure 1. Hence, the light transmission modulation is evidently controlled by GP flake reorientation and relaxation. Thus, electric field-controlled uniform reorientation and relaxation of GP flakes and the subsequent modulation of transmission intensity provide us a promising method for making new polarizer-independent electro-optic flake devices.

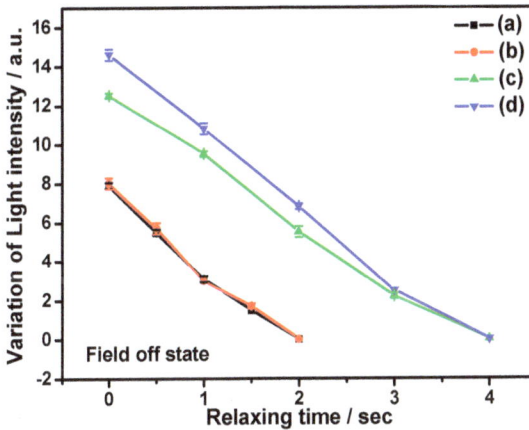

Figure 3. The measured relaxation-induced light intensity variation of representative GP flakes according to Figure 1. The field off state for Case I is 0 s, 0.5 s, 1 s, 1.5 s, and 2 s, and for Case II is 0 s, 1.0 s, 2.0 s, 3.0 s, and 4.0 s, respectively (Case I: flake area and aspect ratio in (**a,b**): 35 μm^2 with an aspect ratio of 2:1 and 85 μm^2 with an aspect ratio of 1:1; Case II: flake area and aspect ratio in (**c,d**): 37 μm^2 with an aspect ratio of 2:1 and 90 μm^2 with an aspect ratio of 1:1). The scattered points represent the experimental data along with measurement error bars, and the lines are guides for the eye. Error bars for respective data points are assigned considering standard deviations obtained in repetitive observations.

Figure 4 shows the NLC host-type and cell thickness-dependent reorientation time of GP flakes with comparable flake morphology under gradually-increasing AC electric field strength from 0 mV/μm to 200.0 mV/μm. It is worth mentioning here that the reported reorientation time has been measured from hundreds of case studies for flakes of several different shapes with aspect ratios from 1:1 to 1:3 and flake area in the range 20 μm^2 to 90 μm^2 while suspended in different NLC materials viz. ZIL4792 and ZIL4535, and variable cell confinement, i.e., cell thicknesses of 44 μm and 70 μm. Hence, we have classified numerous case studies for the reorientation time of GP flakes based on widely different dielectric anisotropy of the NLC host and cell confinement, i.e., the thickness of the used cells. Case I, Case II, and Case III of Figure 4 exhibit the reorientation time results in representative groups obtained under different experimental conditions with low dielectric anisotropy of the NLC host and low cell thickness, high dielectric anisotropy of the NLC host, and low cell thickness, low dielectric anisotropy of the host NLC, and high cell thickness, respectively. Reasonable differences obtained in the reorientation time of GP flakes in the NLC matrix for unique magnitudes of applied electric fields are represented by vertical error bars. Such variations are attributed to the differences in shape, area, and aspect ratio of the GP flakes suspended in the NLC matrix within the abovementioned range, as discussed in our earlier publication [5]. The electric field-dependent reorientation time of GP flakes shows a wide range of variation from 6.5 s to 1.5 s. Reorientation time of GP flakes in the NLC matrix clearly exhibits a decreasing trend with increasing applied electric field strength until saturation for all three cases. The vertical error bars for respective data points are assigned considering the standard deviations obtained through repetitive observations are also found to be significantly suppressed due to the high reproducibility of the recorded reorientation time at significantly higher fields.

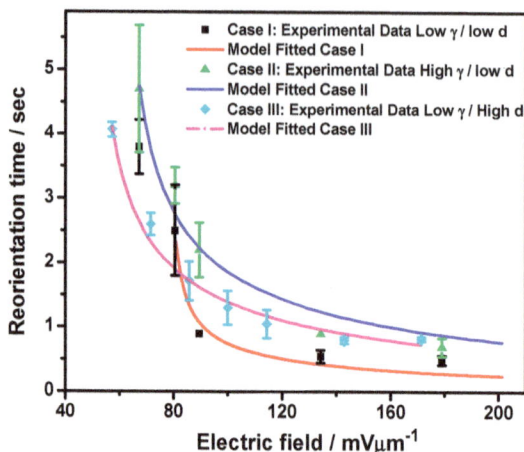

Figure 4. The averaged reorientation time of representative GP flakes with different NLC host-types and cell thicknesses as a function of the electric field (Case I: low γ, low d; Case II: high γ, low d; Case III: low γ, high d). The scattered points represents experimental data, along with measurement error bars and the lines, are model fitted with $Y = P_1/(X^2 + P_2)^{0.5}$ following Equation (4), where P_1 is the numerator of Equation (4), and P_2 is the constant part in the denominator of Equation (4) divided by coefficient of E^2. Error bars for respective data points are assigned considering the standard deviations obtained of repetitive observations.

As the reorientation time of GP flakes in NLC with the field switched on is described in Equation (2), in Case I and Case II it is observed that strong dependence of the NLC host-types in identical cell confinement conditions exists. It is also found that GP flakes in NLC, viz. ZLI-4792 (Case I), having lower dielectric anisotropy ($\Delta\varepsilon = 5.3$), lower rotational viscosity ($\gamma = 132$ mPa·s), and higher splay elastic constant ($K_{11} = 13.2 \times 10^{-12}$ N) reorient quicker than that in NLC, viz. ZLI-4535 (Case II), having higher dielectric anisotropy ($\Delta\varepsilon = 14.8$), higher rotational viscosity ($\gamma = 148$ mPa·s), and lower ($K_{11} = 9.3 \times 10^{-12}$ N). The positive dielectric anisotropic NLC host experiences dielectric torque proportionate with the dielectric anisotropy of the NLC host under an oscillatory AC field. As the π-π interaction between the GP surface and NLC molecules is well known [5–8,11–13], NLC molecules adsorbed over the GP surface should induce a proportionate dielectric torque induced over the NLC host by the oscillating field to the suspended GP flakes. In addition, a comparable Maxwell-Wagner polarization is assumed to be induced over the GP surface under similar experimental conditions. Hence, GP flakes suspended in NLC, viz. ZLI-4535, should experience stronger dielectric torque in comparison with GP flakes suspended in NLC, viz. ZLI-4792, and should reorient faster. However, the larger rotational viscosity of ZLI-4535 hinders the reorientation process and GP flakes suspended in ZLI-4792 exhibit faster reorientation time, which agrees with our previous model that the rotational viscosity is one important factor for GP flakes' reorientation time [5].

In order to determine the cell confinement conditions necessary for GP device application we have further examined the reorientation time of GP flakes in NLC, viz. ZLI-4792, with widely different cell thicknesses of ~44 μm (Case I) and 70 μm (Case III). The experimental results clearly suggest GP flakes suspended in thinner cells reorient faster than in thicker cells. The phenomenon might be anticipated from the considerable difference in the anchoring conditions of the NLC molecules in widely different cell thicknesses. The thicker cell (Case III) imposes weaker anchoring over NLC molecules, and dielectric torque-induced reorientation, and subsequent secondary-induced torque over suspended GP flakes remain weaker than its thinner counterpart (Case I and Case II). Hence, dependence of the reorientation time over the nature of NLC host properties is evident.

For practical display applications the relaxation time of the GP device is also indispensable. The relaxation process of GP flakes in the NLC matrix from the field-aligned edge-up state to the planar face-up state after removing the electric field is recorded under an optical microscope. We have defined the relaxation time as the time taken by GP flakes to relax from the edge-up to the face-up configuration. Figure 5 shows the NLC host-type and cell thickness-dependent relaxation time of GP flakes with comparable flake morphology as a function of reorienting the AC electric field strength from 40 mV/μm to 200 mV/μm. Every case study of GP flakes' field-off relaxation is followed by the abovementioned electric field-induced GP flake reorientation. Thus, an equivalent number of case studies in different NLC matrices and cell confinement with identical experimental conditions to the investigation of reorientation time is guaranteed. We have further regrouped the experimental relaxation time results similar to earlier demonstrated reorientation time cases. Hence, Case I, Case II, and Case III of Figure 5 exhibit the relaxation time results in representative groups obtained under different experimental conditions with low rotational viscosity of the NLC host and low cell thickness, high rotational viscosity of the NLC host, and low cell thickness, low rotational viscosity of the NLC host and high cell thickness, respectively. Reasonable differences are obtained in the relaxation time of GP flakes in the NLC matrix for distinctive magnitudes of the applied electric field, represented by vertical error bars derived from the standard deviations obtained in repetitive observations. Such variations are attributed to the differences in shape, area, and aspect ratio of GP flakes suspended in the NLC matrix within the previously-mentioned range, as discussed in our earlier publication [5]. The relaxation time of GP flakes is found to be independent of the applied reorienting field strength, as is the uncertainty in the relaxation time computation. However, relaxation time strongly depends on the confinement condition of the GP flakes in the NLC matrix, as well as the nature of the NLC host material.

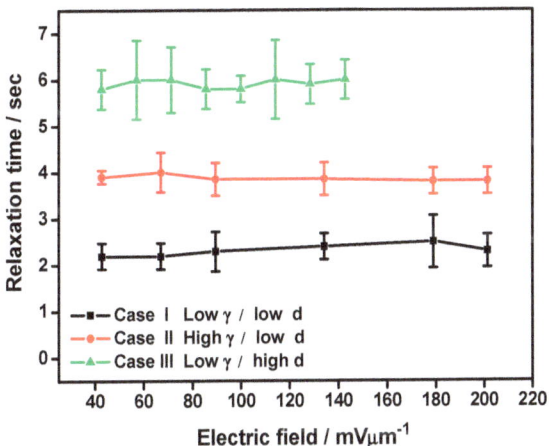

Figure 5. The relationship of the averaged relaxation time of representative GP flakes with different NLC host-types and cell thicknesses after removing the electric field (Case I: low γ, low d; Case II: high γ, low d; and Case III: low γ, high d). The scattered points represent the experimental data along with the measurement error bars, and the lines are guides for the eye. Error bars for respective data points are assigned considering the standard deviations obtained in repetitive observations.

Nanomaterials **2017**, *7*, 250

The GP flakes in NLC, viz. ZLI-4792 (Case I), having lower rotational viscosity (γ = 132 mPa·s) and higher splay elastic constant (K_{11} = 13.2 × 10^{-12} N) relax faster than that in NLC, viz. ZLI-4535 (Case II), having higher rotational viscosity (γ = 148 mPa·s) and lower (K_{11} = 9.3 × 10^{-12} N) in a similar confinement condition. The relaxation process of GP flakes in the NLC matrix is attributed to the relaxation of the NLC host. The π-π electron stacking between the graphene's honeycomb structure and the NLC's benzene rings results in a considerable amount of charge transfer and strong binding of NLC molecules over graphene surface with a binding energy of $U_{anchoring}$ = −2 eV [7,8,13]. Density functional calculations further support the said facts [16]. In relaxation, this strong anchoring energy between NLC molecules and the graphene surface drives the GP flake to rotate following relaxation of the NLC director. However, the rotational viscosity of the NLC matrix hinders the relaxation process and, hence, faster relaxation time has been realized in the NLC host having lower rotational viscosity, viz. ZLI-4792 (Case I).

In addition, the cell thickness-dependent relaxation time of GP flakes with the same NLC host (Case I and Case III) has also been reported in Figure 5. Here, GP flakes' relaxation time in NLC, viz. ZLI-4792, with 44 μm (Case I) has been compared to that in NLC, viz. ZLI-4792, with 70 μm (Case III), where the GP flakes with similar morphology have been tested [5]. The averaged relaxation time in Case I with lower d is found to be quicker than that in Case III. As mentioned earlier, the thinner cell imposes a stronger surface-induced anchoring over confined NLC molecules. Hence, NLC molecules adsorbed over the GP surface will induce stronger relaxing torque onto the GP flake due to the strong coupling of graphene/NLC than its thicker counterpart. Hence, dependence of the relaxation time over cell confinement conditions is evident. This observation is in compliance with the relaxation mechanism of NLC as described elsewhere [9,17–19].

4. Conclusions

We have demonstrated the electric field-induced dynamic reorientation and relaxation processes of GP flakes in an NLC host. The field-induced reorientation time of GP flakes has been effectively reduced by using the suitable physical properties of an NLC host. Our investigation reveals quicker reorientation time of GP flakes in the NLC host with higher K_{11} and lower γ_1. Considering the flake reorientation phenomenon in positive dielectric anisotropy NLCs, the reorientation process depends not only on NLC director reorientation, but also the interfacial Maxwell-Wagner model. At the interface between GP flakes and NLC molecules, strong π-π interaction exists, and the relaxation process is owing to the elasticity-driven rotation of NLC director. The higher K_{11} and lower γ_1 are responsible for higher elastic torque, which matches with the modified rising time equation well in conformity with our previous reported reorientation model. In a field-off state, the relaxation of GP flakes mainly follows the nematic NLC director that mechanically rotates after the field switches off. The GP flakes relax quicker in the NLC host with higher K_{11} and lower γ_1 than its higher counterpart, and with lower d than its thicker counterpart, where the anchoring effect over the NLC director and subsequent secondary induced torque over suspended GP flakes are mainly responsible for this behavior. The reversible reorientation behavior of GP flakes in NLC and the improved response open up an efficient way of dynamically and quickly controlling transmittance properties of electro-optic devices, which makes them interesting candidates for potential applications.

Acknowledgments: This research was supported by Basic Research Laboratory Program (2014R1A4A1008140) through the Ministry of Science, ICT & Future Planning and Polymer Materials Fusion Research Center. W. Tie also thanks to the support from the National Nature Science Foundation of China (61605167), the Science and Technology Research Fund of Henan Provence (162102410093, 172102310478) and XuChang City, and the Key Scientific Research Project of Universities and Colleges in Henan Province (17A430028).

Author Contributions: Seung Hee Lee conceived and designed the experiments; Weiwei Tie, Eun Jeong Shin, Tae Hyung Kim, and MinSu Kim performed the experiments; Seung Hee Lee, Weiwei Tie, Surjya Sarathi Bhattacharyya, and Joong Hee Lee analyzed the data; Seung Hee Lee, Yuanhao Gao and Zhi Zheng contributed reagents/materials/analysis tools; and Weiwei Tie, Surjya Sarathi Bhattacharyya, Seung Hee Lee, and MinSu Kim wrote the paper.

Nanomaterials **2017**, *7*, 250

Conflicts of Interest: The authors declare no conflict of interest.

References

1. Kosc, T.Z.; Marshall, K.L.; Trajkovska-Petkoska, A.; Kimball, E.; Jacobs, S.D. Progress in the development of polymer cholesteric liquid crystal flakes for display applications. *Displays* **2004**, *25*, 171–176. [CrossRef]
2. Trajkovska-Petkoska, A.; Varshneya, R.; Kosc, T.Z.; Marshall, K.L.; Jacobs, S.D. Enhanced electro-optic behavior for shaped polymer cholesteric liquid-crystal flakes made using soft lithography. *Adv. Funct. Mater.* **2005**, *15*, 217–222. [CrossRef]
3. Novoselov, K.S.; Geim, A.K.; Morozov, S.V.; Jiang, D.; Zhang, Y.; Dubonos, S.V.; Grigorieva, I.V.; Firsov, A.A. Electric field effect in atomically thin carbon films. *Science* **2004**, *306*, 666–669. [CrossRef] [PubMed]
4. Meyer, J.C.; Geim, A.K.; Katsnelson, M.I.; Novoselov, K.S.; Booth, T.J.; Roth, S. The structure of suspended graphene sheets. *Nature* **2007**, *446*, 60–63. [CrossRef] [PubMed]
5. Tie, W.W.; Bhattacharyya, S.S.; Lim, Y.J.; Lee, S.W.; Lee, T.H.; Lee, Y.H.; Lee, S.H. Dynamic electro-optic response of graphene/graphitic flakes in nematic liquid crystals. *Opt. Express* **2013**, *21*, 19867–19879. [CrossRef] [PubMed]
6. Al-Zangana, S.; Iliut, M.; Turner, M.; Vijayaraghavan, A.; Dierking, I. Properties of a thermotropic nematic liquid crystal doped with graphene oxide. *Adv. Opt. Mater.* **2016**, *4*, 1541–1548. [CrossRef]
7. Basu, R.; Kinnamon, D.; Garvey, A. Nano-electromechanical rotation of graphene and giant enhancement in dielectric anisotropy in a liquid crystal. *Appl. Phys. Lett.* **2015**, *106*, 201909. [CrossRef]
8. Basu, R.; Iannacchione, G.S. Orientational coupling enhancement in a carbon nanotube dispersed liquid crystal. *Phys. Rev. E* **2010**, *81*, 051705. [CrossRef] [PubMed]
9. Jakeman, E.; Raynes, E.P. Electro-optic response times in liquid crystals. *Phys. Lett. A* **1972**, *39*, 69–70. [CrossRef]
10. Shen, T.Z.; Hong, S.H.; Song, J.K. Electro-optical switching of graphene oxide liquid crystals with an extremely large Kerr coefficient. *Nat. Mater.* **2014**, *13*, 394–399. [CrossRef] [PubMed]
11. Kim, D.W.; Kim, Y.H.; Jeong, H.S.; Jung, H.T. Direct visualization of large-area graphene domains and boundaries by optical birefringence. *Nat. Nanotechnol.* **2011**, *7*, 29–34. [CrossRef] [PubMed]
12. Scalia, G.; Lagerwall, J.P.F.; Haluska, M.; Dettlaff-Weglikowska, U.; Giesselmann, F.; Roth, S. Effect of phenyl rings in liquid crystal molecules on SWCNTs studied by Raman spectroscopy. *Phys. Status Solidi* **2006**, *243*, 3238–3241. [CrossRef]
13. Basu, R.; Garvey, A.; Kinnamon, D. Effects of graphene on electro-optic response and ion-transport in a nematic liquid crystal. *J. Appl. Phys.* **2015**, *117*, 074301. [CrossRef]
14. Jin, M.H.; Jeong, H.K.; Kim, T.H.; So, K.P.; Cui, Y.; Yu, W.J.; Ra, E.J.; Lee, Y.H. Synthesis and systematic characterization of functionalized graphene sheets generated by thermal exfoliation at low temperature. *J. Phys. D Appl. Phys.* **2010**, *43*, 275402. [CrossRef]
15. Trushkevych, O.; Collings, N.; Hasan, T.; Scardaci, V.; Ferrari, A.C.; Wilkinson, T.D.; Crossland, W.A.; Milne, W.I.; Geng, J.; Johnson, B.F.G.; et al. Characterization of carbon nanotube-thermotropic nematic liquid crystal composites. *J. Phys. D Appl. Phys.* **2008**, *41*, 125106. [CrossRef]
16. Park, K.A.; Lee, S.M.; Lee, S.H.; Lee, Y.H. Anchoring a liquid crystal molecule on a single-walled carbon nanotube. *J. Phys. Chem. C* **2007**, *111*, 1620–1624. [CrossRef]
17. Nie, X.Y.; Lu, R.B.; Xianyu, H.Q.; Wu, T.X.; Wu, S.T. Anchoring energy and cell gap effects on liquid crystal response time. *J. Appl. Phys.* **2007**, *101*, 103110. [CrossRef]
18. Lim, Y.J.; Bhattacharyya, S.S.; Tie, W.W.; Park, H.R.; Lee, Y.H.; Lee, S.H. Effects of carbon nanotubes on electro-optic characteristics in vertically aligned liquid crystal display. *Liq. Cryst.* **2013**, *40*, 1202–1208. [CrossRef]
19. Kim, J.H.; Kang, W.S.; Choi, H.S.; Park, K.; Lee, J.H.; Yoon, S.; Yoon, S.; Lee, G.D.; Lee, S.H. Effect of surface anchoring energy on electro-optic characteristics of a fringe-field switching liquid crystal cell. *J. Phys. D Appl. Phys.* **2015**, *48*, 465506. [CrossRef]

nanomaterials

MDPI

Review

Lyotropic Liquid Crystal Phases from Anisotropic Nanomaterials

Ingo Dierking [1,*] **and Shakhawan Al-Zangana** [2]

[1]   School of Physics and Astronomy, University of Manchester, Oxford Road, Manchester M13 9PL, UK
[2]   College of Education, University of Garmian, Kalar 46021, Iraq; shakhawan.al-zangana@hotmail.com
*   Correspondence: ingo.dierking@manchester.ac.uk

Received: 11 August 2017; Accepted: 14 September 2017; Published: 1 October 2017

Abstract: Liquid crystals are an integral part of a mature display technology, also establishing themselves in other applications, such as spatial light modulators, telecommunication technology, photonics, or sensors, just to name a few of the non-display applications. In recent years, there has been an increasing trend to add various nanomaterials to liquid crystals, which is motivated by several aspects of materials development. (i) addition of nanomaterials can change and thus tune the properties of the liquid crystal; (ii) novel functionalities can be added to the liquid crystal; and (iii) the self-organization of the liquid crystalline state can be exploited to template ordered structures or to transfer order onto dispersed nanomaterials. Much of the research effort has been concentrated on thermotropic systems, which change order as a function of temperature. Here we review the other side of the medal, the formation and properties of ordered, anisotropic fluid phases, liquid crystals, by addition of shape-anisotropic nanomaterials to isotropic liquids. Several classes of materials will be discussed, inorganic and mineral liquid crystals, viruses, nanotubes and nanorods, as well as graphene oxide.

Keywords: liquid crystal; lyotropic; inorganic nanoparticle; clay; tobacco mosaic virus (TMV); Deoxyribonucleic acid (DNA); cellulose nanocrystal; nanotube; nanowire; nanorod; graphene; graphene oxide

1. Introduction

Liquid crystals (LC) are a state of matter which is thermodynamically located between the isotropic liquid and the crystalline phase [1,2]. They exhibit flow properties like a liquid and at the same time partially retain the order of a crystal. For this reason, they possess anisotropic physical properties such as their refractive index, dielectric constant, elastic behaviour, or viscosity, just to name a few. But while being partially ordered, LCs also exhibit flow properties like a liquid; they are thus anisotropic fluids. The liquid crystalline state can be brought about via two fundamentally different ways, leading to the two basic classes of LC, thermotropic phases and lyotropic phases. The former is achieved by varying an intensive variable of state, such as temperature or pressure, while the latter is formed through a variation of the concentration of a dopant in an isotropic solvent, often water.

1.1. Thermotropic Liquid Crystals

Thermotropic LCs [3–5] are the ones which are widely known due to their applicational impact in flat screen televisions, laptop and tablet displays, or mobile phones [6]. All these applications rely on the fact that LCs exhibit elastic behaviour and can be addressed via electric or magnetic fields, which changes the orientation of the optic axis, and thus the birefringence. Thermotropic LCs are further distinguished by their degree of order, showing further phase transitions within the temperature regime of the liquid crystalline state. The phase generally observed below the isotropic liquid is called nematic, N, and exhibits solely orientational order of the long axis of rod-like molecules, while disk-like molecules, so called discotic, can show nematic phases as well. The spatial and temporal average of this long axis is called the director, *n*. At lower temperatures smectic phases are also observed, which, in addition to orientational order, also exhibit one- or two-dimensional positional order of the molecules centres of mass. Depending on the degree and nature of order, a whole range of different smectic phases can be distinguished, with the simplest being the smectic A phase with one-dimensional positional order and the director in direction of the smectic layer normal. If the director, which at the same time is the optic axis of the system, is inclined to the layer normal, one speaks of the smectic C phase (see Figure 1).

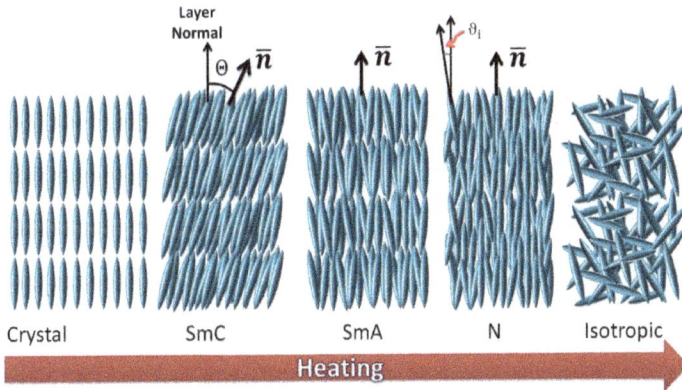

Figure 1. Schematic illustration of different liquid crystal (LC) phases observed on heating from the crystalline state. At first, positional ordering is partially maintained in the smectic phases, SmC and SmA, together with orientational order of the long molecular axis of often rod-shaped molecules. On further heating, positional order is lost at the transition to the nematic phase, which solely exhibits orientational order. Eventually, at the clearing point, all liquid crystalline order is lost and the isotropic liquid is reached. For simplicity, only rod-like molecules are depicted in the figure, but other molecular shapes exist as well, such as disc-like or bent-core materials, which exhibit liquid crystalline behavior.

These phases are called the fluid smectic phases, with hexatic phases and higher ordered phases to follow at even lower temperature [7]. The order thus increases with decreasing temperature, while the symmetry is reduced. An important parameter in the description of LC phases is the orientational order parameter, S_2, which in its simplest description takes the form:

$$S_2 = \langle P_2(cos\vartheta) \rangle = \langle \frac{3cos^2\vartheta - 1}{2} \rangle \qquad (1)$$

where P_2 is the second Legendre polynomial and ϑ the angle between the long axis of an individual molecule and the director. The order parameter changes as a function of temperature, a dependence which is schematically shown in Figure 2.

1.2. Lyotropic Liquid Crystals

Lyotropic LCs [8,9] on the other hand are observed when changing the concentration of a shape- or property anisotropic dispersant in an isotropic solvent. Often, lyotropic phases are observed as a function of concentration of amphiphilic molecules in water or other solvents, as schematically shown in Figure 3. Below the critical micelle concentration, cmc, the amphiphiles are molecularly dispersed in the solvent, but at larger concentrations form micelles, which can be of the spherical, disk or rod-like type, depending on the molecular shape. At even higher concentrations, these micelles aggregate to ordered structures and can form hexagonal, cubic or lamellar phases, also of the inverse type. The observed phase diagrams can be quite complex, as they depend largely on concentration, but also on temperature.

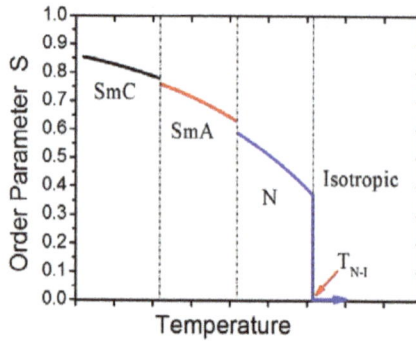

Figure 2. Schematic temperature dependence of the scalar orientational order parameter S. At elevated temperatures in the isotropic phase, it is S = 0. At the clearing temperature $T_{N\text{-}I}$, a first order transition into the nematic phase is observed, accompanied by a discontinuous jump of the order parameter, generally to S ≈ 0.45, which then increases with decreasing temperature to value of about S ≈ 0.6–0.7. Further increases in orientational order are observed at the transitions into smectic phases, albeit much smaller than those between the nematic and the isotropic phase.

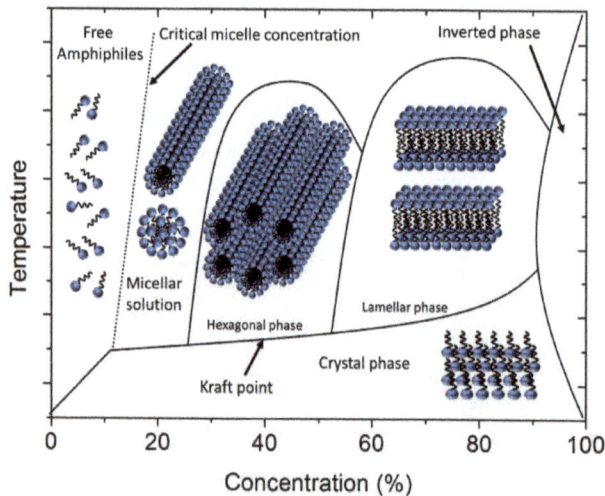

Figure 3. Schematic illustration of the phase diagram of an amphiphilic surfactant in an isotropic solvent, forming lyotropic phases. Crossing the critical micelle concentration, cmc, spherical or cylindrical micelles are formed. At higher surfactant concentrations, these can aggregate to liquid crystalline phases, namely the hexagonal and the lamellar phase, for increasing concentration. Cubic phases, which are not shown in this figure, can occur at different regimes of the phase diagram.

Similarly, dispersions of shape-anisotropic nanoparticles, like nanorods or nanoplates, in isotropic solvents can lead to the formation of often nematic lyotropic LC phases. Spontaneous self-organization is observed, such that the dispersed particles order roughly parallel. The structure is very much reminiscent of its two-dimensional analogue of floating trees on the surface of a lake. Figure 4 shows one of the first photographs of Spirit Lake after the eruption of the volcano Mt. St. Helens in 1980, taken by Col. David K. Wendt, USAF Reserves, one day after the explosive eruption. Whole forests were washed into the lake, where the logs collected to form a "nematic" structure with the logs locally pointing approximately in the same direction, minimizing the free volume. This average direction would then be defined as the director, *n*, in the case of Figure 4 approximately along the diagonal from bottom left to top right.

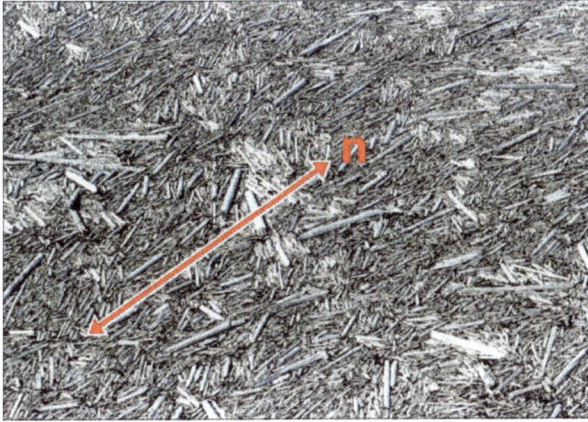

Figure 4. Logs washed into the Spirit Lake after the eruption of the volcano St. Helens in 1980. The photograph was taken from a helicopter by Col. David K. Wendt, USAF Reserves, who was one of the first arriving with a rescue team, one day after the eruption. The logs exhibit the nematic ordering of rigid rods, as proposed by Lars Onsager. (The length of the picture is estimated to approximately 50 m).

In general, the colloidal suspensions of geometrically anisotropic particles can be observed to produce a LC phase above a critical concentration. Orientational order arises from particle anisotropy for an associated critical volume fraction V_{crit} depending on the aspect ratio, $A_R = W/L$, of the particle as $V_{crit} \approx 4/A_R = 4 L/W$, where L is the length and W the width of the nanoparticle. This first theoretical description was reported by Onsager [10]. His theory is based on the fact that when the concentration of particles reaches a certain level, the freedom of the particles is constrained and as a result, the entropy decreases due to overlapping excluded volume of the particles. To compensate for the entropy decrease, the particles start to acquire orientational order. Thus, above a critical concentration orientation order is achieved and a nematic LC phase is formed. First experimental reports go back even further when liquid crystalline behaviour was described for tobacco and tomato mosaic virus (TMV) [11] and vanadium pentoxide (V_2O_5) [12] in the early 20th century. Decades later Mourchid et al., studied colloidal plate-like charged particles and clay particles to report liquid crystalline behaviour [13]. In this review, we will try to summarize some of the most interesting experimental systems and properties of liquid crystalline behaviour from shape-anisotropic nanoparticles, such as inorganic and mineral materials, clays, biological nanoparticles, such as TMV, DNA and cellulose nanocrystals, nanotubes and nanorods, as well as nanoplates and two-dimensional materials, such as graphene, graphene oxide, and reduced graphene oxide.

1.3. Isotropic to Nematic Transition: Maier-Saupe vs. Onsager

Thermotropic LCs are generally described by the Maier-Saupe theory [14–16], which is based exclusively on attractive interactions of the van der Waals type, thus induced dipole-induced dipole interactions. This works very well, because the rigid, polarizable cores of respective mesogens are mainly of the aromatic type, and steric repulsion can largely be ignored. Through a self-consistent field equation, which has to be solved numerically, the Maier-Saupe theory predicts the temperature dependence of the (scalar) order parameter S_2. At a certain temperature a discontinuous, first order transition is observed, at which the order parameter takes a value of $S_2 = 0.43$, continuously increasing with decreasing temperature, to values of the order of $S_2 = 0.7–0.8$ for rather low temperatures. The Maier-Saupe approach does not yield a satisfactory description of lyotropic LCs though, especially those based on colloidal particles as they are discussed in this publication.

Already a decade before the work of Maier and Saupe, Onsager [10] formulated a theory which was able to describe the occurrence of a nematic state in colloidal suspensions, and which was largely the opposite of the approach that Maier and Saupe took later on. Onsager's theory starts from the assumption that no two particles can occupy the same space, all interactions between suspended colloids are in fact repulsive, ignoring any attractive van der Waals forces. Repulsive interactions can be steric, thus based on excluded volume, or they can be of electrostatic nature. Onsager in fact accounted for electrostatic repulsion as approximating it through an increase in particle size, which was later detailed more correctly [17,18]. Further simplifications that were made in the first instance, but relaxed later on (see [18] and references therein), were monodispersity and the use of rigid rod particles of length L and diameter D. Thus, the aspect ratio D/L plays a paramount role in Onsager's approach. The description is purely based on maximizing the entropy. The phase behaviour of such a colloidal suspension is found by the minimization of the free energy F = U − TS, where the internal energy U = 0 for pure steric repulsion, T is the temperature and S the entropy. There are two competing effects: decreasing the excluded volume increases the free volume and thus the translational entropy of the particles. This on the other hand implies a reduction of the mixing entropy. When the particle concentration is small the mixing entropy term dominates the phase behaviour and the isotropic phase is maintained. According to Onsager this is the case for particle volume fractions of $\Psi_{iso} < 3.3$ D/L. At this corresponding particle concentration, a first order transition is observed into the two-phase region of coexisting isotropic and nematic phase. This two-phase region terminates into the nematic phase at a particle volume fraction of $\Psi_n > 4.5$ D/L. A complete phase diagram for rigid rod cylinders from computer simulation by Bolhuis and Frenkel [19] is depicted in Figure 5, which includes not only the isotropic and the nematic phase, but also smectic ordering and colloidal crystals.

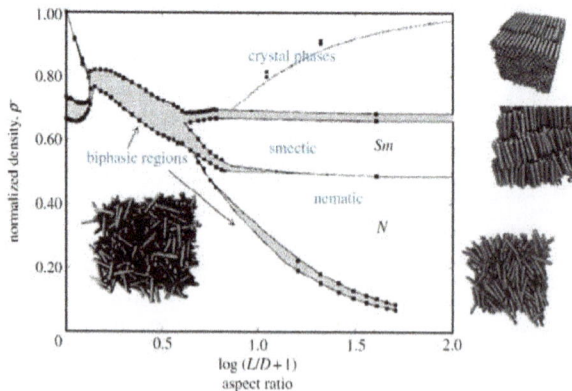

Figure 5. Predicted phase diagram from computer simulations of a rigid rod system, showing the isotropic phase at low aspect ratios and low concentrations, then an Iso + N biphasic region, before a nematic phase is established. Eventually, also smectic and crystalline phases are observed. (Reproduced by permission from ref. [19]).

The Onsager model leads to very high values of the orientational order parameter of $S_2 \approx 0.8$ at the beginning of the transition into the nematic state, to $S_2 \approx 0.95$ at larger nanoparticle concentrations. These values are much larger than those predicted for thermotropic nematic phases by the Maier-Saupe theory, but are indeed observed for lyotropic phases from anisotropic colloidal particles, as we will see below in an example of the TMV. It is believed that the Maier-Saupe model is more applicable to systems that are only slightly compressible, thus only show a small change of density at the transition and smaller orientational order parameters, while the Onsager model is most appropriately applied to dilute suspensions of particles, showing large changes in density and orientational order parameter at the transition to the liquid crystalline state.

2. Lyotropic Phases from Nanomaterials

Much effort has been invested into the study of nanoparticles dispersed in LCs. These systems are mainly studied for their possibility to tune the liquid crystalline properties, such as threshold voltage V_{th}, response times τ, viscosity η, dielectric anisotropy $\Delta\varepsilon$, refractive index n, or the birefringence Δn, which are of importance for applications, especially in the area of LC displays (LCDs). Nanoparticles employed in such dispersions are often based on inorganic or mineral materials [20,21] such as dielectric and ferroelectric particles like TiO_2 [22] and $Sn_2P_2S_6$ [23] or $BaTiO_3$ [24,25], gold nanoparticles [26] or carbon based nanotubes [27,28] and graphene oxide [29]. However, also other nanotubes, like ZnO [30] or semiconducting CdSe [31] have been employed to add additional functionality to the LC matrix. Not surprisingly, many of the shape-anisotropic dopants can also form LC phases by themselves, through dispersion in isotropic solvents. The shape-anisotropy may be provided through the nanoparticles being rod- or tube-like, disk-like or existing as sheet-like materials. These are then representing lyotropic LC phases, similar to the ones formed by amphiphilic molecules dispersed at relatively large concentrations above the cmc in a solvent, often water.

Having shortly outlined above the main predictions and descriptions expected from the theory first devised by Lars Onsager in 1949 [10], we can step back in time for a few decades to discuss the first experimentally observed liquid crystalline systems of this kind.

2.1. Inorganic and Mineral Liquid Crystals

LC phases from anisotropic, inorganic or mineral crystallites in a solvent, often water, represent a sol with particles of colloidal size, i.e., particles with at least one dimension smaller than approximately one micrometer. The first studies of such materials go back to 1902 when sols of FeOOH were shown to become birefringent under the application of a magnetic field, today known as the Majorana effect [32]. A few years later, in 1915–1916, Freundlich [12,33] demonstrated on colloidal dispersions of vanadium pentoxide, V_2O_5, the occurrence of birefringence induced by flow alignment of anisotropic crystallites, as well as by applied electric field, and concluded that the mechanism of both effects was the same; application of a force to align the long axis of the anisotropic particles leads to an induced birefringence. Removal of the external stimuli causes a thermal relaxation back to an isotropic distribution of particles without any birefringence.

It has been reported repeatedly that freshly prepared V_2O_5 sols show no birefringence at early times, while only later on, after days, sometimes weeks, anisotropic, birefringent regions develop in the form of tactoids [34]. These are domains with orientationally ordered rod-like particles as shown schematically in Figure 6a. Their shape is clearly different from the circular domains of thermotropic nematics forming at the isotropic to nematic transition, which exhibits round domains due to the minimization of surface tension. Tactoids, which are shown in microscopic observation in Figure 6b for V_2O_5, on the other hand exhibit two tips which may be rounded off if flow or electric and magnetic field application is involved. The growth of tactoids has also been reported for other inorganic sols [20] like H_2WO_4 and FeOOH, but also for other systems, like chromonic LCs [35,36] or TMV [37]. For inorganic or mineral sols, nematic liquid crystalline behaviour with purely orientational order has been observed, but also smectic type phases with additional ordering in layers can be found.

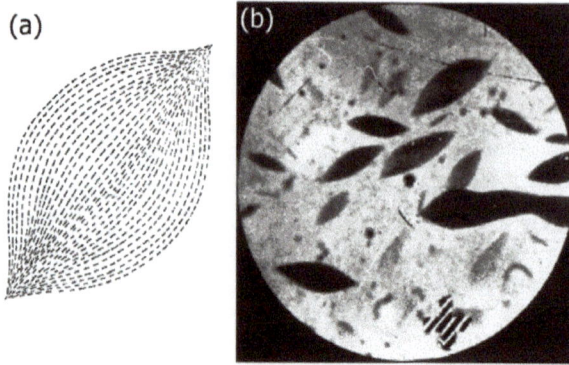

Figure 6. (a) schematic illustration of the director/particle field within a nematic tactoid; and (b) microscopic photograph of the same for the inorganic LC vanadium pentoxide, V_2O_5. (Reproduced by permission from ref. [34]).

The classic inorganic LC is vanadium pentoxide V_2O_5, which forms a nematic lyotropic phase. The phase formation is strongly dependent on the preparation conditions of V_2O_5, which shows small elongated crystallites or fibre like ribbons. This is related to the aging of virgin preparations over time periods of hours or even days, depending on temperature, concentration and electrolyte addition. The aging process, thus the formation process of a nematic phase from the sol, increases in speed for large crystallite concentrations, higher temperatures, and increased electrolyte addition. The particle length then increases at practically constant width of approximately 10 nm from several nanometers to a few micrometers, which is accompanied by a sol-gel transition [38]. Electric field experiments in the nematic phase indicate a negative dielectric anisotropy, $\Delta\varepsilon < 0$. A respective temperature-concentration stability diagram is shown for V_2O_5 in Figure 7.

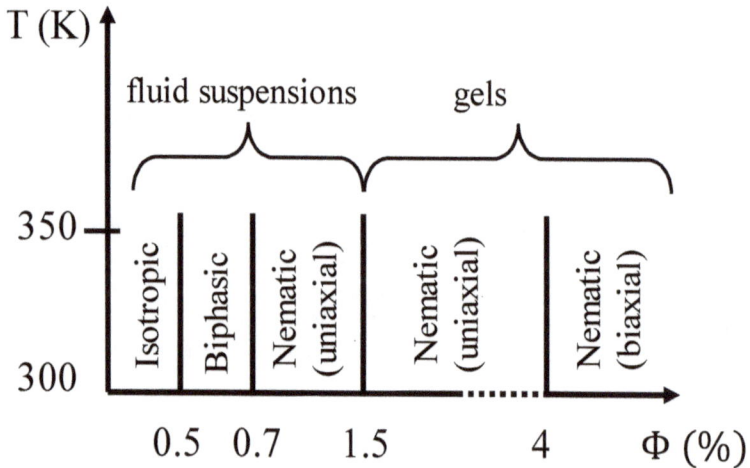

Figure 7. Summary of the phase behaviour of V_2O_5 in dependence on the volume fraction of the dispersed inorganic crystallites, as determined by Nuclear Magnetic Resonance, NMR. (Reproduced by permission from ref. [39]).

Similar behaviour as for vanadium pentoxide is observed for aluminium oxyhydroxide, AlOOH, with tactoids of nematic order forming. When these tactoids join, a Schlieren texture with typical disclinations is developed, as shown in Figure 8a [40]. Another nanomaterial to mention is uranyl fluoride UO_2F_2. This exhibits a nematic phase in a solution of acetone and heavy water, as demonstrated by nuclear magnetic resonance (NMR) experiments [41].

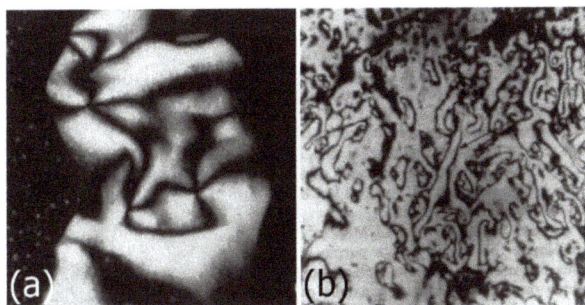

Figure 8. (**a**) Nematic Schlieren texture of AlOOH (reproduced by permission from ref. [40]); and (**b**) nematic thread-like texture of $Li_2Mo_6Se_6$ (reproduced by permission from ref. [42]), scale unknown.

A general class of inorganic nematic LCs is $M_2Mo_6X_6$ with the metal M = Li, Na, K from the alkalimetal group 1, and X = Se, Te from the chalcogens group 16. Also here we observe crystallite lengths of a few micrometers and the formation of Schlieren textures or thread-like textures [42], as shown in Figure 8b, clearly identifying nematic behaviour, in this case with N-methylformamide as solvent. The phase separation into a nematic and an isotropic component is observed after several hours to months.

As mentioned above, also the formation of smectic phases can be observed. This has been demonstrated for example for FeOOH by the microscopic observation of step textures indicating smectic layering [43], and for tungstic acid H_2WO_4 ($WO_3 \cdot H_2O$). A detailed overview about preparation methods and conditions of elongated inorganic particles and their LC structures and phases can be found in a review article by Sonin [20].

2.2. Clay Based Liquid Crystals

Clays are obviously a very closely related topic, as they are aluminosilicates, rock-forming minerals. They generally exist in powder form with layered structures made of plates or platelets. This structure is also the reason why they easily swell in the presence of water or other isotropic solvents [44], which are situated between the sheets. Similar to the hard-rod model of Onsager and its later variants, such a model can also be formulated for hard disks, and computer simulations by Veerman and Frenkel indicate a stable nematic and a columnar phase [45].

2.2.1. Bentonite

Already Langmuir [46] in 1938 observed that a bentonite clay suspension phase separates into an isotropic region and an anisotropic, birefringent region, which in fact turns out to be a gel. Only much later it was realized that at very low concentrations a real LC phase and not a gel could be produced. Bentonite clay is used in a number of diverse applications from food additives to cosmetics, facial masks, and nutritional products, all the way to cat litter and drilling mud. The general application of the liquid crystalline phase may be found in the ordering of inorganic nanosheets for possible future applications in sensors or energy storage. Hybrids with dyes could be used for photosensitization, or in optical materials for plasmonic nanostructures.

2.2.2. Laponite

Laponite, a synthetic clay used in personal care products to modify rheological properties, or as gelator in construction work and artware, also exhibits a similar behaviour. Some exemplary textures of concentrated solutions are shown in Figure 9 for both materials [47]. These are typical nematic textures from disk-like platelets.

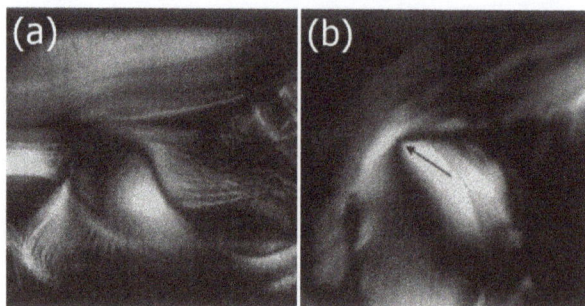

Figure 9. Nematic texture of different clays, (**a**) bentonite and (**b**) laponite (reproduced by permission from ref. [47]). The image width is approximately 1 mm for (**a**) and 500 μm for (**b**).

Applications of the liquid crystalline phase could again be found in the ordering of inorganic nanosheets by templating LC order for sensors and optical materiuals, or in the application of cosmetics.

2.2.3. Imogolite

Imogolite is a natural hydrated aluminium silicate found in volcanic ash, which can be dispersed in water under acidic conditions. The observed phase separation between isotropic and anisotropic, birefringent liquid is practically temperature independent [48,49]. Imogolite forms a tubular structure and exhibits a texture reminiscent of a cholesteric fingerprint organisation, demonstrated in Figure 10.

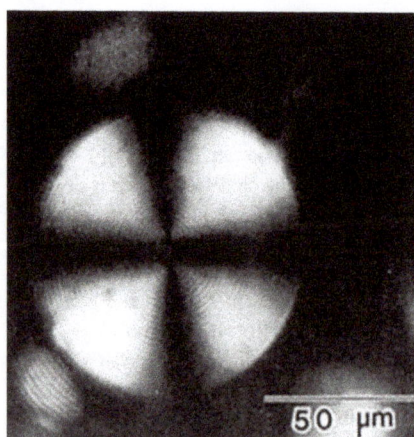

Figure 10. So-called "fingerprint" texture of imogolite, exhibiting an equidistant line pattern, which is somewhat reminiscent of a cholesteric fingerprint texture. Nevertheless, imogolite does not exhibit any chiral constituents, so that the stripe pattern is not indicative of a cholesteric phase. (Reproduced by permission from ref. [49]).

It does not seem to be quite clear why a chiral texture is observed, while the tubes do not seem to exhibit any chirality. Also, the equidistant striped pattern disappears for increasing concentrations of imogolite. Possibly, the texture is related to a banded texture, as it can also be observed for some nematic gels formed from molecular ribbons.

2.3. Biological Nanoparticles

Biological or natural fibres, in fact all biological anisotropic nanoparticles, are potential candidates for exhibiting lyotropic liquid crystalline behaviour. For example, cellulose and its derivatives shows cholesteric lyotropic phases in many isotropic solvents [50], and so do cellulose nanocrystals in water [51]. Fibrous or filamentous proteins for example of the collagen family, or muscle proteins like actin can for LC phases. The variety of liquid crystalline behaviour of nucleic acidsDeoxyribonucleic acid (DNA) or ribonuceic acid (RNA) is multiple, and also related rod-like structures, such as the TMV, has long been shown to exhibit mesophases.

2.3.1. Tobacco Mosaic Virus (TMV)

The TMV, is a single-stranded RNA virus which affects mainly tobacco plants, but also various other plants, visible by a discolouration of the leaves in a mosaic like pattern. Back in the 1930s it was presumably the first ever virus to be discovered. The TMV can be seen as a natural prototype of a ridged rod. It is very straight and rod-like in structure with a constant diameter of approximately 18 nm and an often uniform length of about 300 nm, thus an aspect ratio of the order of $D/L \approx 15$ (see Figure 11a). It is thus ideally suited as a test system for the Onsager theory.

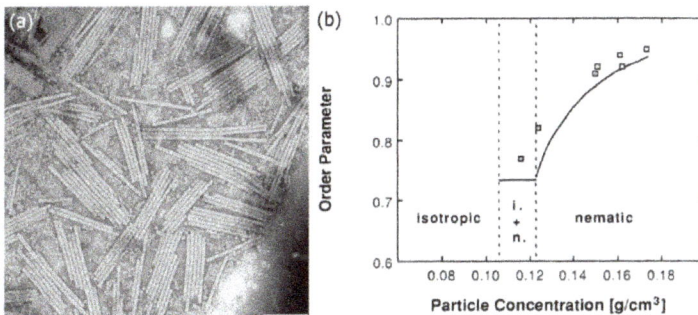

Figure 11. (a) Electron microscopic photograph of tobacco mosaic viruses, TMV, indicating an aspect ratio of approximately 15, and a relatively monodisperse length distribution. The scale bar indicates 0.2 μm. Part (b) depicts the concentration dependence of the orientational order parameter S, as determined by small angle x-ray scattering, SAXS. The order parameter is zero in the isotropic liquid, and increases from about $S \approx 0.7$ to $S \approx 0.95$ through the biphasic region and into the regime of the nematic phase at large concentrations. (part (a) is reproduced from wikimedia commons, with no author name supplied, while part (b) is reproduced by permission from ref. [52]).

And indeed, in 1936 a first publication [11] reported liquid crystalline behaviour for TMVs at certain concentrations, through the growth of tactoids and with corresponding x-ray investigations (figure with textures). Most likely independently derived, a similar result was reported by Rischkov and Smirnova [53] about five years later. More detailed small angle X-ray scattering (SAXS) experiments were carried out by Oldenbourg and co-workers [52] who produced a small angle diffraction pattern for a magnetic field aligned nematic phase, which allowed the determination of the orientational order distribution function, and thus the scalar order parameter S. The latter indicated a transition from isotropic through a two-phase region into a nematic phase with increasing TMV concentration.

The order parameter in the nematic phase changed from about S = 0.75 at the transition to S = 0.95 for high concentrations, which is in accordance with Onsager's predictions (Figure 11b).

Fraden and co-workers [54] performed a thorough investigation of TMV solutions, measuring the birefringence as a function of concentration, temperature, ionic strength and polydispersity. They observed the appearance of spatial and angular local order for increasing concentration until an isotropic to nematic phase transition is accomplished via an Iso-N two-phase regime. From their measurements, together with a modified Onsager approach, they concluded that the nematic phase stability of TMV suspensions is mainly caused by electrostatic repulsion, rather than attractive (van der Waals) forces between the TMV rods. This indicates a transition due to excluded volume effects as generally predicted by Onsager, whose theory ignores attractive forces all together and is purely based on (steric) repulsion interactions. The complete phase diagram of the tobacco mosaic virus was later predicted from theory and simulations by Graf and Löwen [55], who did not only describe the isotropic to nematic transition with its two-phase region, but also further transitions into smectic phases and colloidal crystal phases.

Tobacco mosaic viruses in the nematic state have also been suggested for the design of silica-TMV mesostructures and nanoparticles, i.e., using the TMV as a template in the synthesis of inorganic frameworks with ordered porosity in order to produce more complicated structures. Fowler et al. [56] describe a method where ordered TMVs in the nematic arrangement are silicated and then thermally removed via biodegradation. This led to silica structures with hexagonally ordered nano-channels of a diameter of approximately 20 nm. The authors also synthesized silica nanoparticles with radially arranged nano-channels. A range of further approaches to use TMV LCs as templates for the controlled synthesis of ordered structures have since been pointed out (see references [57,58] and references therein). It is worth noting, that while the TMV produces nematic ordering, also cholesteric [59] and smectic phases [60] have been observed for different virus suspensions.

2.3.2. DNA

Since the identification of the double helix structure of DNA by Watson and Crick in 1953, based on the X-ray information provided by Gosling and Franklin, this biopolymer has attracted much interest, not only as the carrier of genetic information, but also as a molecule to probe elasticity, as a component in hybrid materials or in bio-nanotechnology. The DNA structure is composed of two helical biopolymers coiled around each other, bound together by hydrogen bonds. Along its length, the structure of DNA is quite flexible and dynamic. The double helix is about 2 nm wide and has a pitch of 3.4 nm, while a DNA molecule can contain millions of base pairs and can have a length in the order of millimetres, and even tens of millimetres.

The fact that DNA can exhibit liquid crystalline phases has been realized quite some time ago, and there has been an excellent review article of the earlier work until about twenty years ago by Livolant and Leforestier [61]. In 1988, Strzelecka et al. [62] reported on the multiple LC phases of DNA at higher concentrations. Through NMR line width experiments on solutions of DNA fragments of approximately 50 nm in average length, corresponding to 146 base pairs, the isotropic, cholesteric and smectic-like phases, together with their respective two-phase regions iso + chol and chol + smectic were identified as a function of increasing DNA concentration. Furthermore, typical textures were shown, similar to the ones of Figure 12. Already one year later, it was shown by Livolant et al. [63] that the high concentration phase is of the columnar type, and later, by the same authors, it was further demonstrated by electron microscopy of typical double twist cylinders, that the often called "pre-cholesteric" phase was in fact a Blue Phase [64]. The structure and evolution of the liquid crystalline phases of DNA was also confirmed by polarization sensitive two-photon fluorescence microscopy on respectively labelled DNA molecules.

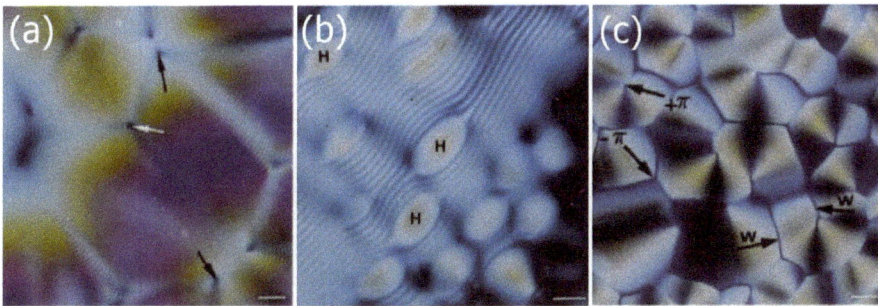

Figure 12. Typical textures observed for the lyotropic phases of relatively long DNA with increasing concentration. (**a**) Cholesteric fingerprint texture with equidistant line pattern due to the helical superstructure of the phase. The distance between two dark lines is equal to identity period of half the pitch, P/2; (**b**) At the transition from the cholesteric to the columnar hexagonal phase; and (**c**) within the fully developed columnar hexagonal phase. The scale bars are 10 μm. (Reproduced by permission from ref. [61]).

In recent years, the attention of LC forming DNA has slightly shifted towards rather short, more controllable and defined molecules. According to Onsager theory only volume fractions of $\Psi > \Psi_{Iso-N} \approx 4D/L$ should be able to exhibit a nematic phase. Molecules with aspect ratios $L/D < 4$ should not show LC phase formation for any concentrations. Nonetheless, Nakata et al. [65] have demonstrated the 6–20 base pair DNA can in fact exhibit cholesteric and columnar phases via end-to-end adhesion and stacking of oligomers into polydisperse, rod-shaped, semi-rigid aggregates, which then act like colloidal particles forming LC phases. This mechanism of self-assembly of short DNA duplexes to form building blocks for cholesteric LCs was detailed later on in subsequent publications [66]. In the original work on short DNA molecules, fully complimentary sequences were employed, which was then extended by Zanchetta et al. [66,67] to partially overlapping sequences and even to LC ordering in systems with a large amount of randomness provided by random DNA sequences, when Bellini et al. [68] discussed liquid crystalline behaviour observed between the isotropic phase of short oligomers and isotropic gels of long random DNA strands. Very recently, this work on very short DNA has been extended to systems with only four base pairs to still show liquid crystalline behaviour via aggregation [69]. An interesting effect for short DNA LCs has been observed for varying concentration, as right handed DNA macromolecules can produce left handed cholesteric structures at low concentration and right handed ones at larger concentrations, passing a structurally non-chiral nematic state as a function of concentration [70]. This is very similar to the temperature induced twist inversion phenomena observed in thermotropic LCs [71–76] where in the lyotropic systems the concentrations mimics the role of temperature as the variable of state.

DNA finds its applications in LC research and possible future technology for example in the use as chiral dopants to control the pitch of cholesteric phases [77], as biosensors [78] and even as alignment layers for LC applications [79].

2.3.3. Cellulose Nanocrystals

Cellulose is composed of polysaccharides, linear chains of hundreds to thousands of sugar units. It is a natural polymer which occurs in abundance in nature and has one of the most inexhaustible supplies. Its attractiveness for materials and composites lies in the fact that cellulose is regenerative, easily biodegradable, and optical properties have been studied in great detail. It has long been known that cellulose derivatives in suitable solvents form lyotropic LCs, often with a rather short pitch in the visible range of the spectrum [50,80]. Their phase behaviour and optical properties have been studies in great detail, also with respect to cellulose-based derivatives. Cellulose nanocrystals as hard rod systems seem to have attracted increasing interest only over recent years.

Like many of the other hard-rod nano-materials we have discussed above, also cellulose nanocrystals form lyotropic LCs in accordance with the predictions of Onsager's model. Also in this case tactoids may be observed in microscopy of cellulose nanocrystal/water systems. Only here, we are dealing with chiral constituent molecules, such that a chiral nematic or cholesteric phase is observed in contrast to the nematic phases. The cellulose nanocrystals have an average dimension of about 100 nm in length, 25 nm in width and approximately 10 nm in height (Figure 13a). They can thus be seen as lath-like nanoparticles [81]. In the lyotropic cholesteric phase, these nanocrystals orient with their long axis parallel to an average direction, which changes with a continuous twist when proceeding in the direction of the short axis, perpendicular to the long axis. Thus a helical superstructure is formed. For increasing concentration of cellulose nanocrystals, the volume fraction of the anisotropic phase increases, as expected (Figure 13b,d). The observed pitch is generally of the order of 10–20 µm, decreasing with increasing cellulose nanocrystal concentration and increasing with nanocrystal length or aspect ratio [82]. The helical superstructure and pitch is clearly manifested also in the observed textures between crossed polarizers, as an equidistant pattern of dark lines, called a fingerprint texture (Figure 13c).

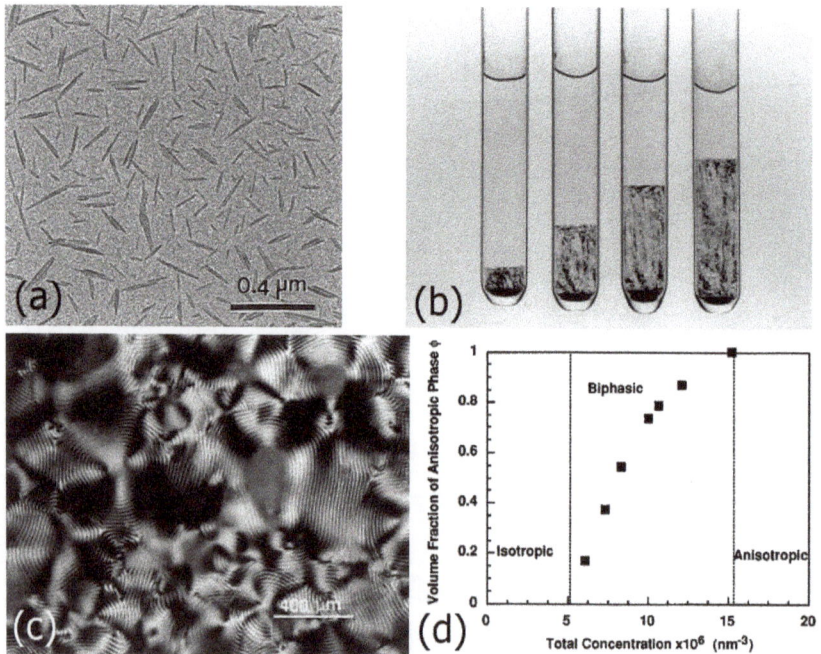

Figure 13. Summary of the basic lyotropic liquid crystalline behaviour of cellulose nanocrystals. (**a**) The nanocrystals of cellulose are composed of chiral polymers and exhibit a length of about 100 nm and lateral dimensions between 10–20 nm, thus aspect ratios in the order of 10; (**b**) For concentration up to about 3% the isotropic phase is observed, which changes to a biphasic region in which the liquid crystalline volume fraction increases with increasing concentration until at about 10–14% a completely anisotropic phase is observed; (**c**) Since the building blocks are chiral, cellulose nanocrystal LCs exhibit a cholesteric phase, as demonstrated by the fingerprint texture. The helical pitch is in the order of 10–20 µm; (**d**) Volume fraction of the anisotropic phase as a concentration of cellulose nanocrystals. (The different parts of the figure were reproduced by permission from ref. [83]).

The general phase behaviour of cellulose nanocrystals in water has been discussed on several occasions [84–88] and results appear to be non-contradictory: below about 3 vol % the solution is completely isotropic. It is followed by a two-phase region of isotropic + LC in the regime of 3–10 vol % nanocrystalline cellulose, and reaches a completely liquid crystalline state at 10–14 vol %. Above 14 vol % a gel is formed [84]. In the two-phase region the anisotropic volume fraction increases with increasing cellulose nanocrystal concentration.

One of the most prominent material parameters of lyotropic cellulose nanocrystal dispersions is the viscosity, which generally increases for increasing concentration and aspect ratio [89]. Also studies relating to the ionic strength have been reported [90], and further the influences of temperature and humidity [91] for dried cellulose nanocrystal films obtained from lyotropic LCs.

Such materials are produced as multifunctional thin films for applications, for example in varying the wavelength of selective reflection across the film diameter, a property which is due to a varying concentration of cellulose nanocrystals [92]. Also the production of plasmonic films of cellulose nanocrystal cholesterics incorporating gold [93,94] or silver [95] nano-rods has been reported. Such composite films display tuneable chiroptic properties.

2.3.4. Active Liquid Crystals

At this point it is worthwhile to also mention active LCs as an emerging topic of pronounced interest. In general, active matter [96,97] resembles a system composed of many active constituents, each of which consumes energy or converts one form of energy to motion or the exertion of a mechanical force. They are therefore intrinsically non-equilibrium systems. Examples are found in a wide variety of soft matter and biological systems, such as swarms of insects, flocks of birds, school of fish, or closer to the topic of this review, bacteria or microtubules. Systems are often of biological origin, but more lately also synthetically derived. They generally show dynamic self-organization and self-propellation. Active LCs have recently become a topic of much increasing interest [98].

In addition, as liquid crystalline systems, active matter is a non-equilibrium system, like cells for example, assemblies of many molecular units working cooperatively to undergo processes like motion, cell division or replication. These systems are actively driven, thus consume energy, which has to be provided from the surrounding. They can thus not be described by equilibrium statistics. The group of Dogic [99] have demonstrated an interesting example of active matter by the use of stretchable microtubule bundles. They showed that active matter can be hierarchically assembled to mimic LCs, but also polymer gels and emulsions by forming an active percolation network at not too small microtubule concentrations. The demonstrated active LCs form the typical s = ±1/2 defects, indicating nematic order with defects that show spatiotemporal dynamics. If one balances the rates of defect creation and defect annihilation, one can achieve steady-state streaming dynamics, which continues over prolonged time scales. This is a behaviour qualitatively different from non-active nematics, as for the latter the defect dynamics follows particular scaling laws for creation [100] and annihilation [101,102] of topological defects. The group of Lavrentovich [103] used a somewhat different active liquid crystal medium. They dispersed motile bacteria, bacillus subtilis, in a liquid crystalline host and demonstrated that the nematic topological defects can be used to command active matter. By employing a variety of different director fields, they showed that the bacteria senses differences in director field deformation. It was observed, that for pure splay and pure bend deformations the bacteria motion is bipolar, with an equal probability distribution for motion along the director field in either direction. This was different for mixed splay-bend regions, where the motion becomes unipolar, directed towards the positive defects and avoiding negative ones. Lavrentovich et al. thus directed the motion of bacteria by the use of defect patterns, and therefore exerting a directing influence on the otherwise chaotic motion. It is very likely that active liquid crystalline systems will become a direction of research where many interesting fundamental aspects are to be discovered, with a high likeliness of future applications in the areas of biotechnology and medicine.

2.4. Liquid Crystals from Nanotubes and Nanorods

The largest part of the literature and thus experimental investigations, are related to the dispersion of nanotubes within an already existing LC phase [104–106]. This host phase can be nematic, cholesteric or smectic, and already possess a physical functionality, as for example in the form of SmC* ferroelectric liquid crystals (FLC). The aim is to transfer the anisotropic order of the LC onto the dispersed anisotropic particles, the LC acting as a template [27,107]. Due to the properties of the dispersed nanotubes or nanorods, this adds functionality to the dispersion, for example in the form of a switchable conductivity [28]. Orientation of nanotubes and changes of physical properties can also be observed in ferroelectric LCs [108], discotics [109,110], and lyotropic phases [111–114]. Nevertheless, here we will concentrate on the opposite phenomena, the formation of lyotropic LC phases through the addition of nanotubes and nanorods to an isotropic solvent [115–117].

2.4.1. Nanotubes

The possible occurrence of liquid crystalline order was first predicted by Somoza et al. [118] who analysed two limiting approaches theoretically: (i) purely attractive van der Waals interactions between the nanotubes, which led to the formation of nematic and columnar phases for increasing concentration of nanotubes; (ii) solely hard-core repulsion, which led to the formation of nematic and smectic A phases for increasing concentration. The isotropic to nematic transition was found to depend on the length of the nanotubes; increasing with increasing nanotube length. Experimentally, it appears that ultra-sonication is of vital importance to de-bundle the nanotubes, increase tube solubility and lead to the observation of lyotropic behaviour. The nematic phase grows in the form of tactoids with an order parameter increasing from about S ~0.3 to S ~0.5 which increases with increasing time of sonication [119,120], as depicted in Figure 14. It should also be noted that a predicted smectic A phase has not been observed experimentally so far, which can most likely be attributed to the polydispersity of the nanotubes. All in all, the nanotube lyotropic phase formation is quite similar to that of TMVs or DNA.

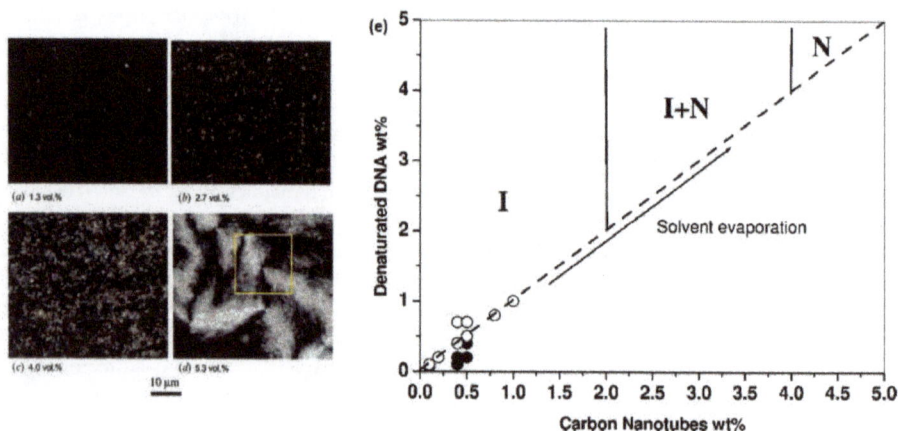

Figure 14. (a–d) formation of the lyotropic nematic phase of multiwall nanotubes, multiwall nanotubes (MWNT), in water, for increasing concentration through the biphasic region. At approximately 5% by volume, the nematic phase is fully developed as evidenced by a typical Schlieren texture. (Reproduced by permission from ref. [117]); (e) Similar results are obtained for DNA functionalized nanotubes. (Reproduced by permission from ref. [121]).

First experimental evidence for lyotropic nanotube LCs was presented shortly after their prediction, by Song et al. [116,117] for a multiwall nanotube (MWNT) in water system. To enhance the solubility without the need for employing a surfactant, the nanotubes were functionallized with COOH before dispersion in water. The transition from isotropic to the nematic state was observed at a nanotube loading of approximately 1 vol %, with a two-phase region between 1–4 vol %. Above this concentration, a purely nematic state was found [115]. The two-phase region is somewhat wider than that predicted by Onsager, which again can be attributed to the large polydispersity of the MWNTs. Windle and co-workers [122] demonstrated that the longer, straighter nanotubes accumulate in the nematic phase of the dispersion, while impurities, which are always present in nanotube systems, as well as short tubes, accumulate in the isotropic liquid. Badaire followed a similar approach for single-wall nanotubes (SWNT), but instead of covalent functionalization, denatured DNA was adsorbed on the walls of the tubes [115]. The dispersion in water is then facilitated via electrostatic repulsion, as the denatured DNA is charged. This is used to compensate the attractive van der Waals interactions between the nanotubes, and implies that below a certain coating concentration the dispersion remains isotropic. Above the critical coating concentration, a nematic phase is observed above 4 wt % SWNTs, with a two-phase region between 2–4 wt % (see Figure 14e). Electrostatic repulsion to disperse the single-wall nanotubes was also used by Rai et al. [121] for nanotube LCs without the functionalization with chemical groups or decoration with DNA. In this case though, a strong acid had to be chosen as the isotropic solvent, which led to protonation of the tube walls and thus electrostatic repulsion and better tube dispersion.

2.4.2. Nanorods and Nanowires

It appears that it is generally hard to obtain large scale uniformly oriented samples of nanotube based lyotropic LCs. This is probably closely related to the largely unavoidable polydispersity of the systems under investigation. It is likely that a more successful approach may be found in the use of nanorods, which can be produced with a much better monodispersity and where the nanoparticles are straight and less flexible, i.e., behave more like an ideal system in terms of the Onsager description. Systems with dispersed nanorods have been investigated, but again, mainly with respect to dispersions in an already existing (thermotropic) LC. Here the self-organization of the LC is exploited to self-assemble nanorods, to provide added functionality or tuning of physical properties. An example are gold nanorods, LC modified gold particles and gold nanorod LCs [26,123–126], which enhance the anisotropy of the conductivity, the dielectric constant, and the elastic behaviour.

Nanorods of ZnO

Zinc oxide, ZnO, is generally produced as a white powder for the use in many materials and applications, such as paints, plastics, glass, ceramics, food products and mainly in the rubber industry, where it is employed in the vulcanization process of rubber. It is a wide band-gap semiconductor of the II-VI group and its uses in the electronics industry are in thin-film transistors, light emitting diodes, and as transparent electrodes for liquid crystal displays.

Mostly, ZnO nanoparticles, and other metallic and metal oxide nanoparticles, are incorporated into already existing thermotropic or lyotropic phases, rather than being used to generate the LC behaviour [30]. Lamellar, cubic and hexagonal lyotropic phases have also been reported to be used as a reaction medium in which nanoparticles are synthesised [127]. Reports on the formation of lyotropic LCs from ZnO nanoparticles are comparatively scarce.

In the form of single crystal semiconductor nanowires ZnO assembles into lyotropic nematic phases in organic and aqueous solvents. The formation of the LC phase follows that predicted by Onsager, and outlined above, where below certain ZnO nanowire concentrations an isotropic phase is formed, which at higher concentrations becomes a two-phase region and eventually at another, still higher concentration, transforms into a lyotropic nematic phase [128]. For the demonstration of such liquid crystalline behaviour, high aspect ratio nanowires were employed, suitably surface-functionalised by molecules containing sulphur, an alkyl spacer and headgroups such as H or COOH. In the nematic state a nicely developed Schlieren texture can be observed, as shown in

Figure 15. On drying thin films from the lyotropic phase, the ZnO nanowires may act like a template of the director field [128], imaging typical s = ±1 and s = ±1/2 defects of the lyotropic nematic phase, similar to polymer stabilized LCs with thermotropic nematic phases [129] (see Figure 15).

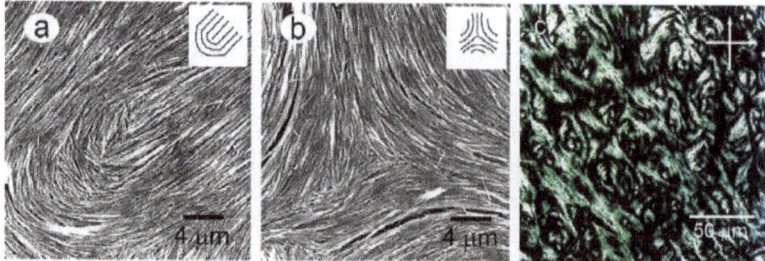

Figure 15. Dried films of ZnO lyotropic nematic phases can be used to image defects of strength (**a**) s = +1/2 and (**b**) s = −1/2; Part (**c**) shows a Schlieren texture of a fully developed lyotropic nematic phase from ZnO nanowires. (Reproduced by permission from ref. [128]).

The same group of authors also went one step further in the functionalization of ZnO nanowires, by doping with cobalt Co and manganese Mn, to introduce magnetic properties [130]. Also here, high aspect ratio, surface functionalized nanowires were used, and magnetic reorientations of the ZnO director field demonstrated.

Nanorods of TiO$_2$

Titanium oxide finds its applications in the food industry, as sunscreen, and especially as a white pigment in paper, plastics and paints. In nature it is known as rutile, anastase and brookite, differing in their crystal structure. TiO$_2$ nanorods and nanowires are generally produced through a conversion of anastase. Also for TiO$_2$ there are reports where lyotropic LC phases are used in the synthesis of nanomaterials, where self-assembled lamellar, spherical and rod-like structures may be observed [131]. Reports of TiO$_2$ nanowires being used to generate lyotropic phases are scarce [132]. One such report describes a two-stage assembly process in the formation of a lyotropic nematic phase, by first forming a primary structure, such as ribbons, which then in a second self-assembly step through an increase in concentration may form a nematic and lamellar lyotropic LC.

CdSe Semiconductor Nanorods

Cadmium selenide nanorods are semiconductors, typically of length ~40 nm and width ~6 nm. This means they have an aspect ratio for which one can well expect the formation of orientational order as it is observed for nematic phases [133]. Due to the fact that these nanoparticles can also be produced with an excellent monodispersity, one even has the opportunity to possibly detect smectic ordering, i.e., the formation of at least one dimensional positional order [134].

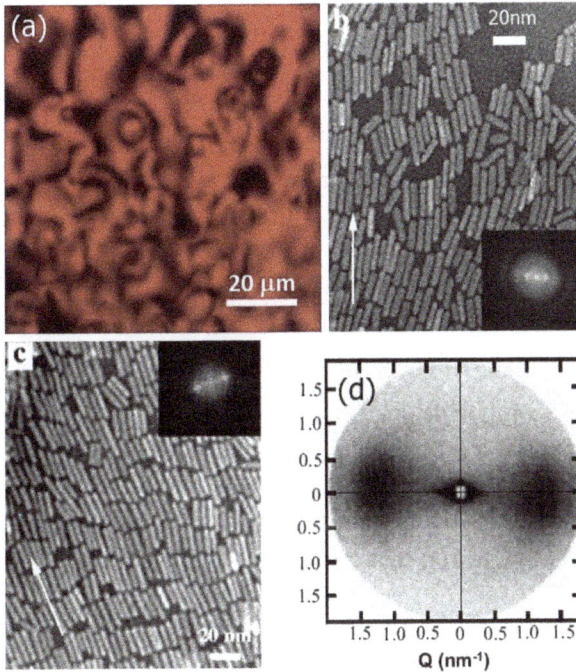

Figure 16. CdSe nanorods exhibit typical nematic Schlieren textures, as shown in part (**a**) of the figure (reproduced by permission from ref. [133]). The electron micrographs of parts (**b**) and (**c**) indicate nematic and smectic ordering, thus orientational and additional one-dimensional positional order, respectively. The insets show the corresponding Fourier transforms. (Reproduced by permission from ref. [135]); (**d**) depicts the SAXS picture of the nematic phase, which also clearly evidences orientational order. (Reproduced by permission from ref. [134]).

CdSe nanorods show indeed a pronounced appearance of lyotropic nematic phases in the presence of organic solvents [134], as shown in the distinct Schlieren textures of Figure 16a, where s = 1/2 and s = 1 disclinations are observable. Also small angle X-ray experiments on oriented samples nicely present evidence for nematic ordering (Figure 16d), while transmission electron microscopy (TEM) reveals not only a nematic structure of nanorods, but also positional order for higher concentrations (Figure 16b,c). Liquid crystalline self-assembly of nanorods has been reviewed recently by Thorkelsson et al. [136]. Not only have semiconducting nanorods been investigated for the formation of lyotropic LCs, but also within LC templates [31].

2.5. Liquid Crystals from Nanoplates

Just like the disk-like colloidal structures of for example clay particles, also other materials of that shape can exhibit very stable lyotropic LC phases as a function of particle concentration. One of the most prominent examples are the derivatives of graphene [137–141]. Graphene has attracted much attention over the recent years due to its promising properties in terms of elastic modulus and conductivity while only exhibiting flakes of the nanometer to micrometer size which are only a carbon monomolecular thick.

2.5.1. Graphene

In terms of liquid crystalline behaviour graphene itself is actually not the material of choice, due to its poor solubility and dispensability in isotropic solvents. This has been tested for a large variety of solvents with varying polarity [142] and it appears that solubility is slightly increased for increasing dielectric constant. Nevertheless, overall the solubility of graphene in any solvent is very small and concentrations to observe lyotropic liquid crystalline behaviour are not easily achieved, not even with prolonged ultrasonication to avoid aggregation and coagulation.

A possible way forward are protonated graphenes. The formation of a lyotropic liquid crystalline phase formed by graphene in chlorosulphonic acid was first reported by Pasquali and co-workers [143] in 2010. A nematic texture was observed to indicate liquid crystallinity. The structure of the nematic phase is similar to that of a discotic nematic phase, with the director being normal to the plane of the graphene sheets. Despite the principle demonstration of liquid crystalline behaviour, processing of these systems is obviously not desirable, and systems with more environmentally friendly solvents and better solubility need to be found.

2.5.2. Graphene Oxide

This is the case with graphene oxide (GO), which represents a form of graphene decorated with hydroxyl, carboxyl and epoxide groups. This makes it easily dissolvable in water and other solvents. A further advantage of graphene oxide is the fact that in contrast to graphene, it is readily available in large quantities at a very reasonable price. As first demonstrated by Kim and co-workers [144] and Xu and Gao [145] in 2011, GO in water or organic solvents forms a nematic phase above a certain threshold concentration, with typical textures observed, as shown in Figure 17a for increasing GO concentration. As common, a two-phase behaviour is observed as demonstrated also in Figure 17b for three different graphene oxide sources. The formed phase is very stable with respect to temperature, up until the boiling point of the solvent.

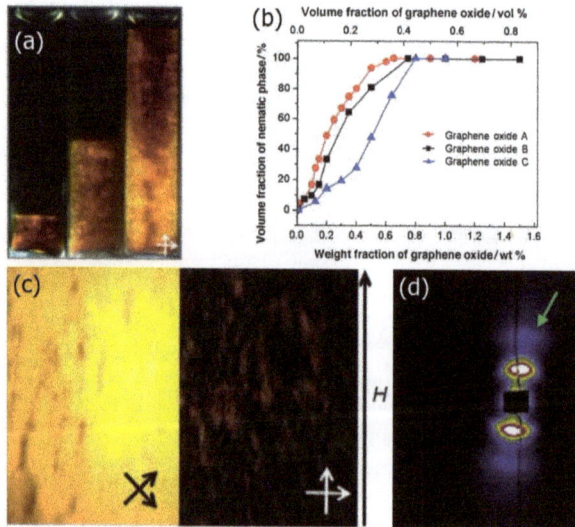

Figure 17. (a) Qualitative illustration of the increasing LC volume fraction for increasing graphene oxide, GO, concentration; Part (b) quantifies this behaviour for three graphene oxide samples from different sources. The difference in quantitative behaviour as the biphasic concentration regime is passed, is due to a variation of polydispersity and graphene flake size among other influences; (c) Application of magnetic fields can be used to uniformly orient the lyotropic nematic phase of graphene oxide, which is evidenced by rotation of the sample between crossed polarizers; (d) Also with small angle X-ray diffraction one can demonstrate orientational order of the director, obtained in capillaries. (Parts (a–c) are reproduced by permission from ref. [144], while part (d) is reproduced by permission from ref. [145]).

Xu and Gao [145] actually claim that the phase they observe can be described by the model of the twist grain boundary (TGB) phase, where blocks of smectic layers are rotated with respect to each other, while the grain boundaries between blocks are arrays of screw dislocations. The rotation of the blocks will eventually lead to a helical superstructure, which can be commensurate or incommensurate. They attribute their interpretation to the observed weak layering by small angle X-ray scattering (Figure 17d) in combination with LC texture observation and cryo-Scanning Electron Microscopy. This appears to be a point of controversy, because the formation of a TGB-like phase requires the presence of chirality, which is absent in the studied system, as neither the graphene oxide, nor the solvent are chiral. The observed textures also appear different than the common fingerprint textures observed for chiral nematic or cholesteric LCs, without a clear periodicity, appearing more like textures observed in shear banding. In addition, the fact that the graphene oxide sheets exhibit a large polydispersity makes it less likely to form a TGB structure. At this point the detailed structure of the observed phase does not seem to be quite clear, and possibly further investigations will be needed. Nevertheless, it is without doubt that the observed aqueous graphene oxide suspensions exhibit liquid crystalline behaviour.

It should further be pointed out, that the actual phase appearance or in fact possibly the diagram slightly depends on the average size of the GO flakes, the polydispersity, the dielectric constant of the solvent and confinement conditions [146]. The liquid crystalline phase is formed at lower concentrations for larger GO flake sizes, it is observed more easily for solvents with an increased dielectric constant, such as water, and it is somewhat suppressed or not observable for more confined geometries. This is most likely due to the very strong planar anchoring of the graphene oxide sheets to the bounding glass substrates, which produces pseudo-isotropic behaviour. The fact that liquid crystalline behaviour of GO can be observed at much lower concentrations if the flakes exhibit a larger size was also observed by Dan et al. [147]. Furthermore, the LC formation in dependence of different organic solvents has been discussed by Jalili et al. [148]. At last, an interesting scenario can be observed if dispersing graphene oxide in a thermotropic nematic. Increasing the temperature above the clearing point, converts the host LC into an isotropic phase, which can then in combination with the GO act as a solvent to form a lyotropic nematic phase [29]. One can thus observe the transition between a thermotropic and a lyotropic nematic phase, which can be shown by dielectric spectroscopy, but is not observable in differential scanning calorimetry, thus apparently not connected to a latent heat [149].

Graphene oxide LCs can be oriented by magnetic field application, as shown in Figure 17c with the corresponding small angle X-ray scattering image showing the typical pattern of orientational order (Figure 17d). Under the confinement of LC sandwich cells, graphene oxide nematic can be oriented between untreated glass plates or in channels, such that the GO plane lies parallel to the substrates [146]. The director therefore is oriented normal to the substrate plane, and the sample can be rotated between well oriented bright and dark states between crossed polarizers (see Figure 18).

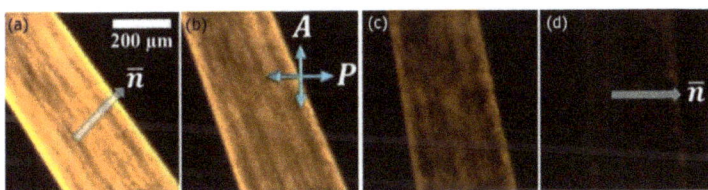

Figure 18. Confinement in channels of plain, untreated glass, can also provide a simple mechanism of orientation for the lyotropic nematic phase of graphene oxide. The transmission of the LC changes with a periodicity of 90° when rotated between crossed polarizes. It is brightest when the director is oriented at 45° to either of the polarizers (**a**) and darkest, when it is parallel to either polarizer A or P (**d**); In between, the transmission continuously varies (**b,c**). (Reproduced by permission from ref. [146]).

Song et al. [150] have demonstrated that application of an AC electric field to a lyotropic graphene oxide nematic LC can result in electro-optic switching, based on the Kerr effect, with a very large Kerr coefficient. This effect can also be used to orient graphene oxide sheets [151].

2.5.3. Reduced Graphene Oxide

Heating graphene oxide above approximately 165 °C thermally reduced GO to rGO, which results in a partial recovery of graphene properties, especially the electronic ones, but at the cost of solubility, which in turn increases the tendency for aggregation and coagulation, making it more difficult to obtain liquid crystalline behaviour. This can be compensated by employing surfactants to stabilize the rGO flakes, as demonstrated by Poulin et al. [152]. One may thus partially maintain the favourable electronic properties of graphene, while additionally being able to exploit the self-organization due to liquid crystallinity.

2.5.4. Other 2D Materials

One could expect that other two-dimensional materials similar to graphene, graphene oxide or reduced graphene oxide, such as boron nitride, indium selenide or gallium selenide, MoS_2, $NbSe_2$, WO_3 or WS_3 can also exhibit lyotropic LC phases at certain concentrations in suitable solvents, especially if these materials occur in single layers. This will then most likely have strong parallels to inorganic LCs and clays.

3. Summary and Outlook

The topic of LCs and nanomaterials has attracted increasing attention over the last years, not only within the LC community, but also more broadly as soft materials in general and model anisotropic colloid systems. An extensive summary of up-to-date knowledge can be found in the two-volume book by Lagerwall and Scalia [153–155]. The three main reasons for this increased interest are (i) nanomaterials in thermotropic LCs can be used to add functionality and tune the properties of the liquid crystalline system; (ii) Phases, especially those of the frustrated type, can be stabilized, and novel materials with anisotropic properties can be created, which spontaneously align shape anisotropic nanoparticles. This can be achieved either through templating liquid crystalline order from thermotropic, as well as lyotropic phases, as well as the formation of lyotropic phases themselves, by nanoparticles ordering in an isotropic solvent; (iii) LC—nanoparticle composites, may these be of the thermotropic or the lyotropic type, allow for the construction of nanotechnology devices in many diverse areas, such as displays, sensors, biological engineering, or even functional clothing. In this review, we have tried to give a broad overview of different lyotropic liquid crystalline systems, based on a variety of anisotropic particles in the colloidal size range. These can be one- or two-dimensional nanomaterials. In both cases, initial investigations on inorganic materials go back for about a century, although they have by far not attracted the attention of their organic, thermotropic counterparts, which is mainly due to the success of the latter in electro-optic and display devices. One of the classic examples of inorganic LCs [20] is vanadium pentoxide, V_2O_5, which dates back to about 1915. While many inorganic LCs are formed by one-dimensional nanoparticles, mineral and clay LCs [44,46] are mostly obtained from plate like, thus two-dimensional particles. The classic examples of biological lyotropic LCs are the TMVs [11] (and other similar viruses), as well as DNA [61]. One material which is located at the borderline between biological one- to two-dimensional crystals, are cellulose nanocrystals [86]. An increasing amount of literature on recent further lyotropic LCs can be found for carbon nanotubes [106,116] (as well as similar nanotubes and nanowires), and graphene oxide [154].

All of the above discussed lyotropic liquid crystalline systems from rods or plates have one feature in common: they all obey the theoretical description initially formulated by Onsager in the 1940s, at least to a large extent, and often even quantitatively. This has also been demonstrated by a variety of computer simulations and experimental work summarized in references [18,19,155].

Nanomaterials **2017**, *7*, 305

Given the synthesis and development of ever new nanomaterials, and the rapid advancement of nanotechnology, it seems to be out of question that lyotropic, anisotropic particle based LCs will play an increasing role of importance in the future. This is mainly due to the fact that many of the functionalities observed and exploited in thermotropic LCs, like electric and magnetic reorientation, and with it a change of birefringence, ferroelectricity or magnetic properties, can now also be observed in lyotropic LCs. The properties of self-assembly, self-organisation, and spontaneous alignment will be beneficial for nanotechnological applications, and the fact that for many of the lyotropic systems, water can be used as a solvent, favours environmentally friendly production mechanisms, which are clearly the way forward for future applications.

Author Contributions: I.D. conceived and wrote this review. S.A.Z. helped with the figures, reproduction permissions, and discussions.

Conflicts of Interest: The authors declare no conflict of interest.

References

1. Collings, P.J. *Liquid Crystals: Nature's Delicate Phase of Matter*; Princeton University Press: Princeton, NJ, USA, 1990.
2. Collings, P.J.; Hird, M. *Introduction to Liquid Crystals Chemistry and Physics*; Taylor & Francis: London, UK; Bristol, PA, USA, 1997.
3. Chandrasekhar, S. *Liquid Crystals*, 2nd ed.; Cambridge University Press: Cambridge, UK, 1992.
4. Singh, S. *Liquid Crystals: Fundamentals*; World Scientific: Hackensack, NJ, USA, 2002.
5. De Gennes, P.G.; Prost, J. *The Physics of Liquid Crystals*, 2nd ed.; Oxford University Press: Oxford, UK; New York, NY, USA, 1993.
6. Lueder, E. *Liquid Crystal Displays: Addressing Schemes and Electro-Optical Effects*; John Wiley & Sons: Chichester, UK, 2001.
7. Dierking, I. *Textures of Liquid Crystals*; Wiley-VCH: Weinheim, Germany, 2003.
8. Petrov, A.G. *The Lyotropic State of Matter: Molecular Physics and Living Matter Physics*; Taylor & Francis: London, UK, 1999.
9. Figueiredo Neto, A.M.; Salinas, S.R.A. *The Physics of Lyotropic Liquid Crystals: Phase Transitions and Structural Properties*; Oxford University Press: New York, NY, USA, 2005.
10. Onsager, L. The effects of shape on the interaction of colloidal particles. *Ann. N. Y. Acad. Sci.* **1949**, *51*, 627–659. [CrossRef]
11. Bawden, F.C.; Pirie, N.W.; Bernal, J.D.; Fankuchen, I. Liquid Crystalline Substances from Virus-infected Plants. *Nature* **1936**, *138*, 1051–1052. [CrossRef]
12. Diesselhorst, H.; Freundlich, H. On the double refraction of vanadine pentoxydsol. *Phys. Z.* **1915**, *16*, 419–425.
13. Mourchid, A.; Delville, A.; Lambard, J.; LeColier, E.; Levitz, P. Phase Diagram of Colloidal Dispersions of Anisotropic Charged Particles: Equilibrium Properties, Structure, and Rheology of Laponite Suspensions. *Langmuir* **1995**, *11*, 1942–1950. [CrossRef]
14. Maier, W.; Saupe, A. Eine einfache molekulare Theorie des nematischen kristallinflüssigen Zustandes. *Zeitschrift für Naturforsch. A* **1958**, *13*, 564–566. [CrossRef]
15. Maier, W.; Saupe, A. Eine einfache molekular-statistische Theorie der nematischen kristallinflüssigen Phase. Teil l1. *Zeitschrift für Naturforsch. A* **1959**, *14*, 882. [CrossRef]
16. Maier, W.; Saupe, A. Eine einfache molekular-statistische Theorie der nematischen kristallinflüssigen Phase. Teil II. *Zeitschrift für Naturforsch. A* **1960**, *15*, 287. [CrossRef]
17. Stroobants, A.; Lekkerkerker, H.N.W.; Odijk, T. Effect of electrostatic interaction on the liquid crystal phase transition in solutions of rodlike polyelectrolytes. *Macromolecules* **1986**, *19*, 2232–2238. [CrossRef]
18. Vroege, G.J.; Lekkerkerker, H.N.W. Phase transitions in lyotropic colloidal and polymer liquid crystals. *Rep. Fmg. Phys.* **1992**, *55*, 1241–1309. [CrossRef]
19. Bolhuis, P.; Frenkel, D. Tracing the phase boundaries of hard spherocylinders. *J. Chem. Phys.* **1997**, *106*, 666–687. [CrossRef]
20. Sonin, A.S. Inorganic lyotropic liquid crystals. *J. Mater. Chem.* **1998**, *8*, 2557–2574. [CrossRef]

21. Gabriel, J.-C.P.; Davidson, P. Mineral Liquid Crystals from Self-Assembly of Anisotropic Nanosystems. *Top. Curr. Chem.* **2003**, *226*, 119–172.

22. Chen, W.-T.; Chen, P.-S.; Chao, C.-Y. Effect of Doped Insulating Nanoparticles on the Electro-Optical Characteristics of Nematic Liquid Crystals. *Jpn. J. Appl. Phys.* **2009**, *48*, 15006. [CrossRef]

23. Reznikov, Y.; Buchnev, O.; Tereshchenko, O.; Reshetnyak, V.; Glushchenko, A.; West, J. Ferroelectric nematic suspension. *Appl. Phys. Lett.* **2003**, *82*, 1917. [CrossRef]

24. Singh, U.B.; Dhar, R.; Dabrowski, R.; Pandey, M.B. Enhanced electro-optical properties of a nematic liquid crystals in presence of BaTiO₃ nanoparticles. *Liq. Cryst.* **2014**, *41*, 953–959. [CrossRef]

25. Al-Zangana, S.; Turner, M.; Dierking, I. A comparison between size dependent paraelectric and ferroelectric BaTiO3 nanoparticle doped nematic and ferroelectric liquid crystals. *J. Appl. Phys.* **2017**, *121*, 85105. [CrossRef]

26. Hegmann, T.; Qi, H.; Marx, V.M. Nanoparticles in Liquid Crystals: Synthesis, Self-Assembly, Defect Formation and Potential Applications. *J. Inorg. Organomet. Polym. Mater.* **2007**, *17*, 483–508. [CrossRef]

27. Dierking, I.; Scalia, G.; Morales, P.; LeClere, D. Aligning and Reorienting Carbon Nanotubes with Nematic Liquid Crystals. *Adv. Mater.* **2004**, *16*, 865–869. [CrossRef]

28. Dierking, I.; Scalia, G.; Morales, P. Liquid crystal–carbon nanotube dispersions. *J. Appl. Phys.* **2005**, *97*, 44309. [CrossRef]

29. Al-Zangana, S.; Iliut, M.; Turner, M.; Vijayaraghavan, A.; Dierking, I. Properties of a Thermotropic Nematic Liquid Crystal Doped with Graphene Oxide. *Adv. Opt. Mater.* **2016**, *4*, 1541–1548. [CrossRef]

30. Saliba, S.; Mingotaud, C.; Kahn, M.L.; Marty, J.-D. Liquid crystalline thermotropic and lyotropic nanohybrids. *Nanoscale* **2013**, *5*, 6641. [CrossRef] [PubMed]

31. Mukhina, M.V.; Danilov, V.V.; Orlova, A.O.; Fedorov, M.V.; Artemyev, M.V.; Baranov, A.V. Electrically controlled polarized photoluminescence of CdSe/ZnS nanorods embedded in a liquid crystal template. *Nanotechnology* **2012**, *23*, 325201. [CrossRef] [PubMed]

32. Majorana, Q. Sur la biréfringence magnétique. *C. R. Acad. Sci.* **1902**, *135*, 159–161.

33. Freundlich, H. Die Doppelbrechung des Vanadinpentoxydsols. *Berichte der Bunsengesellschaft für Phys. Chem.* **1916**, *22*, 27–33.

34. Zocher, H. Spontaneous structure formation in sols; a new kind of anisotropic liquid media. *Anorg. Allg. Chem.* **1925**, *147*, 91–110. [CrossRef]

35. Yi, Y.; Clark, N.A. Orientation of chromonic liquid crystals by topographic linear channels: Multi-stable alignment and tactoid structure. *Liq. Cryst.* **2013**, *40*, 1736–1747. [CrossRef]

36. Kim, Y.-K.; Shiyanovskii, S.V.; Lavrentovich, O.D. Morphogenesis of defects and tactoids during isotropic–nematic phase transition in self-assembled lyotropic chromonic liquid crystals. *J. Phys. Condens. Matter* **2013**, *25*, 404202. [CrossRef] [PubMed]

37. Bernal, J.D.; Fankuchen, I. X-ray and crystallographic studies of plant virus preparations: I. Introduction and preparation of specimens; II. Modes of aggregation of the virus particles. *J. Gen. Physiol.* **1941**, *25*, 111–146. [CrossRef] [PubMed]

38. Davidson, P.; Garreau, A.; Livage, J. Nematic colloidal suspensions of V₂O₅ in water—Or Zocher phases revisited. *Liq. Cryst.* **1994**, *16*, 905–910. [CrossRef]

39. Pelletier, O.; Sotta, P.; Davidson, P. Deuterium Nuclear Magnetic Resonance Study of the Nematic Phase of Vanadium Pentoxide Aqueous Suspensions. *J. Phys. Chem. B* **1999**, *103*, 5427–5433. [CrossRef]

40. Zocher, H.; Török, C. Neuere Beiträge zur Kenntnis der Taktosole. *Kolloid-Zeitschrift* **1960**, *173*, 1–7. [CrossRef]

41. Michaelev, V.A.; Tcherbakov, V.A. *Zh. Obs. Khim.* **1985**, *55*, 1223.

42. Davidson, P.; Gabriel, J.C.; Levelut, A.M.; Batail, P. A New Nematic Suspension Based on All-Inorganic Polymer Rods. *Europhys. Lett.* **1993**, *21*, 317–322. [CrossRef]

43. Zocher, H.; Török, C. Crystals of higher order and their relation to other superphases. *Acta Crystallogr.* **1967**, *22*, 751–755. [CrossRef]

44. Gabriel, J.-C.P.; Davidson, P. New Trends in Colloidal Liquid Crystals Based on Mineral Moieties. *Adv. Mater.* **2000**, *12*, 9–20. [CrossRef]

45. Veerman, J.A.C.; Frenkel, D. Phase behavior of disklike hard-core mesogens. *Phys. Rev. A* **1992**, *45*, 5632–5648. [CrossRef] [PubMed]

46. Langmuir, I. The Role of Attractive and Repulsive Forces in the Formation of Tactoids, Thixotropic Gels, Protein Crystals and Coacervates. *J. Chem. Phys.* **1938**, *6*, 873–896. [CrossRef]

47. Gabriel, J.-C.P.; Sanchez, C.; Davidson, P. Observation of Nematic Liquid-Crystal Textures in Aqueous Gels of Smectite Clays. *J. Phys. Chem.* **1996**, *100*, 11139–11143. [CrossRef]

48. Kajiwara, K.; Donkai, N.; Hiragi, Y.; Inagaki, H. Lyotropic mesophase of imogolite, 1. Effect of polydispersity on phase diagram. *Makromol. Chem.* **1986**, *187*, 2883–2893. [CrossRef]

49. Kajiwara, K.; Donkai, N.; Fujiyoshi, Y.; Inagaki, H. Lyotropic mesophase of imogolite, 2. Microscopic observation of imogolite mesophase. *Makromol. Chemie* **1986**, *187*, 2895–2907. [CrossRef]

50. Zugenmaier, P. *Crystalline Cellulose and Cellulose Derivatives*; Springer: Berlin, Germany, 2010.

51. Lin, N.; Huang, J.; Dufresne, A. Preparation, properties and applications of polysaccharide nanocrystals in advanced functional nanomaterials: A review. *Nanoscale* **2012**, *4*, 3274. [CrossRef] [PubMed]

52. Oldenbourg, R.; Wen, X.; Meyer, R.B.; Caspar, D.L.D. Orientational Distribution Function in Nematic Tobacco-Mosaic-Virus Liquid Crystals Measured by X-Ray Diffraction. *Phys. Rev. Lett.* **1988**, *61*, 1851–1854. [CrossRef] [PubMed]

53. Rischkov, V.L.; Smirnova, V.A. Liquid Crystals of the Virus of the Tobacco Mosaic (Nicotiana virus 1). *Comptes Rendus l'Academie Sci. l'URSS* **1941**, *31*, 930.

54. Fraden, S.; Maret, G.; Caspar, D.L.D. Angular correlations and the isotropic-nematic phase transition in suspensions of tobacco mosaic virus. *Phys. Rev. E* **1993**, *48*, 2816–2837. [CrossRef]

55. Graf, H.; Löwen, H. Phase diagram of tobacco mosaic virus solutions. *Phys. Rev. E* **1999**, *59*, 1932–1942. [CrossRef]

56. Fowler, C.E.; Shenton, W.; Stubbs, G.; Mann, S. Tobacco Mosaic Virus Liquid Crystals as Templates for the Interior Design of Silica Mesophases and Nanoparticles. *Adv. Mater.* **2001**, *13*, 1266–1269. [CrossRef]

57. Flynn, C.E.; Lee, S.-W.; Peelle, B.R.; Belcher, A.M. Viruses as vehicles for growth, organization and assembly of materials. *Acta Mater.* **2003**, *51*, 5867–5880. [CrossRef]

58. Dogic, Z.; Fraden, S. Ordered phases of filamentous viruses. *Curr. Opin. Colloid Interface Sci.* **2006**, *11*, 47–55. [CrossRef]

59. Dogic, Z.; Fraden, S. Cholesteric Phase in Virus Suspensions. *Langmuir* **2000**, *16*, 7820–7824. [CrossRef]

60. Dogic, Z.; Fraden, S. Smectic Phase in a Colloidal Suspension of Semiflexible Virus Particles. *Phys. Rev. Lett.* **1997**, *78*, 2417–2420. [CrossRef]

61. Livolant, F.; Leforestier, A. Condensed phases of DNA: Structures and phase transitions. *Prog. Polym. Sci.* **1996**, *21*, 1115–1164. [CrossRef]

62. Strzelecka, T.E.; Davidson, M.W.; Rill, R.L. Multiple liquid crystal phases of DNA at high concentrations. *Nature* **1988**, *331*, 457–460. [CrossRef] [PubMed]

63. Livolant, F.; Levelut, A.M.; Doucet, J.; Benoit, J.P. The highly concentrated liquid-crystalline phase of DNA is columnar hexagonal. *Nature* **1989**, *339*, 724–726. [CrossRef] [PubMed]

64. Leforstier, A.; Livolant, F. DNA liquid crystalline blue phases. Electron microscopy evidence and biological implications. *Liq. Cryst.* **1994**, *17*, 651–658. [CrossRef]

65. Nakata, M.; Zanchetta, G.; Chapman, B.D.; Jones, C.D.; Cross, J.O.; Pindak, R.; Bellini, T.; Clark, N.A. End-to-End Stacking and Liquid Crystal Condensation of 6- to 20-Base Pair DNA Duplexes. *Science* **2007**, *318*, 1276–1279.

66. Zanchetta, G.; Nakata, M.; Buscaglia, M.; Clark, N.A.; Bellini, T. Liquid crystal ordering of DNA and RNA oligomers with partially overlapping sequences. *J. Phys. Condens. Matter* **2008**, *20*, 494214. [CrossRef]

67. Zanchetta, G.; Nakata, M.; Buscaglia, M.; Bellini, T.; Clark, N.A. Phase separation and liquid crystallization of complementary sequences in mixtures of nanoDNA oligomers. *Proc. Natl. Acad. Sci. USA* **2008**, *105*, 1111–1117. [CrossRef] [PubMed]

68. Bellini, T.; Zanchetta, G.; Fraccia, T.P.; Cerbino, R.; Tsai, E.; Smith, G.P.; Moran, M.J.; Walba, D.M.; Clark, N.A. Liquid crystal self-assembly of random-sequence DNA oligomers. *Proc. Natl. Acad. Sci. USA* **2012**, *109*, 1110–1115. [CrossRef] [PubMed]

69. Fraccia, T.P.; Smith, G.P.; Bethge, L.; Zanchetta, G.; Nava, G.; Klussmann, S.; Clark, N.A.; Bellini, T. Liquid Crystal Ordering and Isotropic Gelation in Solutions of Four-Base-Long DNA Oligomers. *ACS Nano* **2016**, *10*, 8508–8516. [CrossRef] [PubMed]

70. Zanchetta, G.; Giavazzi, F.; Nakata, M.; Buscaglia, M.; Cerbino, R.; Clark, N.A.; Bellini, T. Right-handed double-helix ultrashort DNA yields chiral nematic phases with both right- and left-handed director twist. *Proc. Natl. Acad. Sci. USA* **2010**, *107*, 17497–17502. [CrossRef] [PubMed]

71. Stegemeyer, H.; Siemensmeyer, K.; Sucrow, W.; Appel, L. Liquid Crystalline Norcholesterylesters: Influence of the Axial Methylgroups on the Phase Transitions and the Cholesteric Helix. *Zeitschrift für Naturforsch. A* **1989**, *44*, 1127–1130. [CrossRef]

72. Slaney, A.J.; Nishiyama, I.; Styring, P.; Goodby, J.W. Twist inversion in a cholesteric material containing a single chiral centre. *J. Mater. Chem.* **1992**, *2*, 805–810. [CrossRef]

73. Dierking, I.; Gießelmann, F.; Zugenmaier, P.; Kuczynskit, W.; Lagerwall, S.T.; Stebler, B. Investigations of the structure of a cholesteric phase with a temperature induced helix inversion and of the succeeding SmC* phase in thin liquid crystal cells. *Liq. Cryst.* **1993**, *13*, 45–55. [CrossRef]

74. Styring, P.; Vuijk, J.D.; Nishiyama, I.; Slaney, A.J.; Goodby, J.W. Inversion of chirality-dependent properties in optically active liquid crystals. *J. Mater. Chem.* **1993**, *3*, 399–405. [CrossRef]

75. Dierking, I.; Gießelmann, F.; Zugenmaier, P.; Mohr, K.; Zaschke, H.; Kuczynski, W. The Origin of the Helical Twist Inversion in Single Component Cholesteric Liquid Crystals. *Zeitschrift für Naturforsch. A* **1994**, *49*, 1081–1086. [CrossRef]

76. Dierking, I.; Gießelmann, F.; Zugenmaier, P.; Mohr, K.; Zaschke, H.; Kuczynski, W. New diastereomeric compound with cholesteric twist inversion. *Liq. Cryst.* **1995**, *18*, 443–449. [CrossRef]

77. De Michele, C.; Zanchetta, G.; Bellini, T.; Frezza, E.; Ferrarini, A. Hierarchical Propagation of Chirality through Reversible Polymerization: The Cholesteric Phase of DNA Oligomers. *ACS Macro Lett.* **2016**, *5*, 208–212. [CrossRef]

78. Tan, H.; Yang, S.; Shen, G.; Yu, R.; Wu, Z. Signal-Enhanced Liquid-Crystal DNA Biosensors Based on Enzymatic Metal Deposition. *Angew. Chem. Int. Ed.* **2010**, *49*, 8608–8611. [CrossRef] [PubMed]

79. Nakata, M.; Zanchetta, G.; Buscaglia, M.; Bellini, T.; Clark, N.A. Liquid Crystal Alignment on a Chiral Surface: Interfacial Interaction with Sheared DNA Films. *Langmuir* **2008**, *24*, 10390–10394. [CrossRef] [PubMed]

80. Rojas, O.J. *Cellulose Chemistry and Properties: Fibers, Nanocelluloses and Advanced Materials*; Springer: Cham, Switzerland, 2016.

81. George, J.; Sabapathi, S.N. Cellulose nanocrystals: Synthesis, functional properties, and applications. *Nanotechnol. Sci. Appl.* **2015**, *8*, 45–54. [CrossRef] [PubMed]

82. Gray, D.; Mu, X. Chiral Nematic Structure of Cellulose Nanocrystal Suspensions and Films; Polarized Light and Atomic Force Microscopy. *Materials (Basel)*. **2015**, *8*, 7873–7888. [CrossRef] [PubMed]

83. Dong, X.M.; Kimura, T.; Revol, J.-F.; Gray, D.G. Effects of Ionic Strength on the Isotropic–Chiral Nematic Phase Transition of Suspensions of Cellulose Crystallites. *Langmuir* **1996**, *12*, 2076–2082. [CrossRef]

84. Beck-Candanedo, S.; Roman, M.; Gray, D.G. Effect of Reaction Conditions on the Properties and Behavior of Wood Cellulose Nanocrystal Suspensions. *Biomacromolecules* **2005**, *6*, 1048–1054. [CrossRef] [PubMed]

85. Ureña-Benavides, E.E.; Ao, G.; Davis, V.A.; Kitchens, C.L. Rheology and Phase Behavior of Lyotropic Cellulose Nanocrystal Suspensions. *Macromolecules* **2011**, *44*, 8990–8998. [CrossRef]

86. Lagerwall, J.P.F.; Schütz, C.; Salajkova, M.; Noh, J.; Hyun Park, J.; Scalia, G.; Bergström, L. Cellulose nanocrystal-based materials: From liquid crystal self-assembly and glass formation to multifunctional thin films. *NPG Asia Mater.* **2014**, *6*, e80. [CrossRef]

87. Honorato-Rios, C.; Kuhnhold, A.; Bruckner, J.R.; Dannert, R.; Schilling, T.; Lagerwall, J.P.F. Equilibrium Liquid Crystal Phase Diagrams and Detection of Kinetic Arrest in Cellulose Nanocrystal Suspensions. *Front. Mater.* **2016**, *3*, 21. [CrossRef]

88. Gray, D. Recent Advances in Chiral Nematic Structure and Iridescent Color of Cellulose Nanocrystal Films. *Nanomaterials* **2016**, *6*, 213. [CrossRef] [PubMed]

89. Wu, Q.; Meng, Y.; Wang, S.; Li, Y.; Fu, S.; Ma, L.; Harper, D. Rheological behavior of cellulose nanocrystal suspension: Influence of concentration and aspect ratio. *J. Appl. Polym. Sci.* **2014**, *131*, 40525. [CrossRef]

90. Shafiei-Sabet, S.; Hamad, W.Y.; Hatzikiriakos, S.G. Ionic strength effects on the microstructure and shear rheology of cellulose nanocrystal suspensions. *Cellulose* **2014**, *21*, 3347–3359. [CrossRef]

91. Wu, Q.; Meng, Y.; Concha, K.; Wang, S.; Li, Y.; Ma, L.; Fu, S. Influence of temperature and humidity on nano-mechanical properties of cellulose nanocrystal films made from switchgrass and cotton. *Ind. Crops Prod.* **2013**, *48*, 28–35. [CrossRef]

92. Park, J.H.; Noh, J.; Schütz, C.; Salazar-Alvarez, G.; Scalia, G.; Bergström, L.; Lagerwall, J.P.F. Macroscopic Control of Helix Orientation in Films Dried from Cholesteric Liquid-Crystalline Cellulose Nanocrystal Suspensions. *ChemPhysChem* **2014**, *15*, 1477–1484. [CrossRef] [PubMed]

93. Querejeta-Fernández, A.; Chauve, G.; Methot, M.; Bouchard, J.; Kumacheva, E. Chiral Plasmonic Films Formed by Gold Nanorods and Cellulose Nanocrystals. *J. Am. Chem. Soc.* **2014**, *136*, 4788–4793. [CrossRef] [PubMed]

94. Liu, Q.; Campbell, M.G.; Evans, J.S.; Smalyukh, I.I. Orientationally Ordered Colloidal Co-Dispersions of Gold Nanorods and Cellulose Nanocrystals. *Adv. Mater.* **2014**, *26*, 7178–7184. [CrossRef] [PubMed]

95. Chu, G.; Wang, X.; Chen, T.; Gao, J.; Gai, F.; Wang, Y.; Xu, Y. Optically Tunable Chiral Plasmonic Guest–Host Cellulose Films Weaved with Long-range Ordered Silver Nanowires. *ACS Appl. Mater. Interfaces* **2015**, *7*, 11863–11870. [CrossRef] [PubMed]

96. Menzel, A.M. Tuned, driven, and active soft matter. *Phys. Rep.* **2015**, *554*, 1–45. [CrossRef]

97. Ramaswamy, S. Active matter. *J. Stat. Mech. Theory Exp.* **2017**, *2017*, 54002. [CrossRef]

98. Bukusoglu, E.; Bedolla Pantoja, M.; Mushenheim, P.C.; Wang, X.; Abbott, N.L. Design of Responsive and Active (Soft) Materials Using Liquid Crystals. *Annu. Rev. Chem. Biomol. Eng.* **2016**, *7*, 163–196. [CrossRef] [PubMed]

99. Sanchez, T.; Chen, D.T.N.; DeCamp, S.J.; Heymann, M.; Dogic, Z. Spontaneous motion in hierarchically assembled active matter. *Nature* **2012**, *491*, 431–434. [CrossRef] [PubMed]

100. Fowler, N.; Dierking, D.I. Kibble-Zurek Scaling during Defect Formation in a Nematic Liquid Crystal. *ChemPhysChem* **2017**, *18*, 812–816. [CrossRef] [PubMed]

101. Dierking, I.; Marshall, O.; Wright, J.; Bulleid, N. Annihilation dynamics of umbilical defects in nematic liquid crystals under applied electric fields. *Phys. Rev. E* **2005**, *71*, 61709. [CrossRef] [PubMed]

102. Dierking, I.; Ravnik, M.; Lark, E.; Healey, J.; Alexander, G.P.; Yeomans, J.M. Anisotropy in the annihilation dynamics of umbilic defects in nematic liquid crystals. *Phys. Rev. E* **2012**, *85*, 21703. [CrossRef] [PubMed]

103. Peng, C.; Turiv, T.; Guo, Y.; Wei, Q.-H.; Lavrentovich, O.D. Command of active matter by topological defects and patterns. *Science* **2016**, *354*, 882–885. [CrossRef] [PubMed]

104. Zakri, C. Carbon nanotubes and liquid crystalline phases. *Liq. Cryst. Today* **2007**, *16*, 1–11. [CrossRef]

105. Lagerwall, J.P.F.; Scalia, G. Carbon nanotubes in liquid crystals. *J. Mater. Chem.* **2008**, *18*, 2890–2898. [CrossRef]

106. Yadav, S.P.; Singh, S. Carbon nanotube dispersion in nematic liquid crystals: An overview. *Prog. Mater. Sci.* **2016**, *80*, 38–76. [CrossRef]

107. Lynch, M.D.; Patrick, D.L. Organizing Carbon Nanotubes with Liquid Crystals. *Nano Lett.* **2002**, *2*, 1197–1201. [CrossRef]

108. Yakemseva, M.; Dierking, I.; Kapernaum, N.; Usoltseva, N.; Giesselmann, F. Dispersions of multi-wall carbon nanotubes in ferroelectric liquid crystals. *Eur. Phys. J. E* **2014**, *37*, 7. [CrossRef] [PubMed]

109. Kumar, S.; Bisoyi, H.K. Aligned Carbon Nanotubes in the Supramolecular Order of Discotic Liquid Crystals. *Angew. Chem. Int. Ed.* **2007**, *46*, 1501–1503. [CrossRef] [PubMed]

110. Bisoyi, H.K.; Kumar, S. Carbon nanotubes in triphenylene and rufigallol-based room temperature monomeric and polymeric discotic liquid crystals. *J. Mater. Chem.* **2008**, *18*, 3032. [CrossRef]

111. Lagerwall, J.P.F.; Scalia, G.; Haluska, M.; Dettlaff-Weglikowska, U.; Giesselmann, F.; Roth, S. Simultaneous alignment and dispersion of carbon nanotubes with lyotropic liquid crystals. *Phys. Status Solidi* **2006**, *243*, 3046–3049. [CrossRef]

112. Lagerwall, J.; Scalia, G.; Haluska, M.; Dettlaff-Weglikowska, U.; Roth, S.; Giesselmann, F. Nanotube Alignment Using Lyotropic Liquid Crystals. *Adv. Mater.* **2007**, *19*, 359–364. [CrossRef]

113. Jiang, W.; Yu, B.; Liu, W.; Hao, J. Carbon Nanotubes Incorporated within Lyotropic Hexagonal Liquid Crystal Formed in Room-Temperature Ionic Liquids. *Langmuir* **2007**, *23*, 8549–8553. [CrossRef] [PubMed]

114. Scalia, G.; von Bühler, C.; Hägele, C.; Roth, S.; Giesselmann, F.; Lagerwall, J.P.F. Spontaneous macroscopic carbon nanotube alignment via colloidal suspension in hexagonal columnar lyotropic liquid crystals. *Soft Matter* **2008**, *4*, 570–576. [CrossRef]

115. Badaire, S.; Zakri, C.; Maugey, M.; Derré, A.; Barisci, J.N.; Wallace, G.; Poulin, P. Liquid Crystals of DNA-Stabilized Carbon Nanotubes. *Adv. Mater.* **2005**, *17*, 1673–1676. [CrossRef]

116. Song, W. Nematic Liquid Crystallinity of Multiwall Carbon Nanotubes. *Science* **2003**, *302*, 1363. [CrossRef] [PubMed]

117. Song, W.; Windle, A.H. Isotropic–Nematic Phase Transition of Dispersions of Multiwall Carbon Nanotubes. *Macromolecules* **2005**, *38*, 6181–6188. [CrossRef]

118. Somoza, A.M.; Sagui, C.; Roland, C. Liquid-crystal phases of capped carbon nanotubes. *Phys. Rev. B* **2001**, *63*, 81403. [CrossRef]

119. Puech, N.; Blanc, C.; Grelet, E.; Zamora-Ledezma, C.; Maugey, M.; Zakri, C.; Anglaret, E.; Poulin, P. Highly Ordered Carbon Nanotube Nematic Liquid Crystals. *J. Phys. Chem. C* **2011**, *115*, 3272–3278. [CrossRef]

120. Zakri, C.; Blanc, C.; Grelet, E.; Zamora-Ledezma, C.; Puech, N.; Anglaret, E.; Poulin, P. Liquid crystals of carbon nanotubes and graphene. *Philos. Trans. R. Soc. A Math. Phys. Eng. Sci.* **2013**, *371*, 20120499. [CrossRef] [PubMed]

121. Rai, P.K.; Pinnick, R.A.; Parra-Vasquez, A.N.G.; Davis, V.A.; Schmidt, H.K.; Hauge, R.H.; Smalley, R.E.; Pasquali, M. Isotropic–Nematic Phase Transition of Single-Walled Carbon Nanotubes in Strong Acids. *J. Am. Chem. Soc.* **2006**, *128*, 591–595. [CrossRef] [PubMed]

122. Song, W.; Windle, A.H. Size-Dependence and Elasticity of Liquid-Crystalline Multiwalled Carbon Nanotubes. *Adv. Mater.* **2008**, *20*, 3149–3154. [CrossRef]

123. Sharma, V.; Park, K.; Srinivasarao, M. Colloidal dispersion of gold nanorods: Historical background, optical properties, seed-mediated synthesis, shape separation and self-assembly. *Mater. Sci. Eng. R Rep.* **2009**, *65*, 1–38. [CrossRef]

124. Stamatoiu, O.; Mirzaei, J.; Feng, X.; Hegmann, T. Nanoparticles in Liquid Crystals and Liquid Crystalline Nanoparticles. *Top. Curr. Chem.* **2012**, *318*, 331–393. [PubMed]

125. Liu, Q.; Cui, Y.; Gardner, D.; Li, X.; He, S.; Smalyukh, I.I. Self-Alignment of Plasmonic Gold Nanorods in Reconfigurable Anisotropic Fluids for Tunable Bulk Metamaterial Applications. *Nano Lett.* **2010**, *10*, 1347–1353. [CrossRef] [PubMed]

126. Umadevi, S.; Feng, X.; Hegmann, T. Large Area Self-Assembly of Nematic Liquid-Crystal-Functionalized Gold Nanorods. *Adv. Funct. Mater.* **2013**, *23*, 1393–1403. [CrossRef]

127. Saliba, S.; Davidson, P.; Impéror-Clerc, M.; Mingotaud, C.; Kahn, M.L.; Marty, J.-D. Facile direct synthesis of ZnO nanoparticles within lyotropic liquid crystals: towards organized hybrid materials. *J. Mater. Chem.* **2011**, *21*, 18191. [CrossRef]

128. Zhang, S.; Majewski, P.W.; Keskar, G.; Pfefferle, L.D.; Osuji, C.O. Lyotropic Self-Assembly of High-Aspect-Ratio Semiconductor Nanowires of Single-Crystal ZnO. *Langmuir* **2011**, *27*, 11616–11621. [CrossRef] [PubMed]

129. Dierking, I.; Archer, P. Imaging liquid crystal defects. *RSC Adv.* **2013**, *3*, 26433–26437. [CrossRef]

130. Zhang, S.; Pelligra, C.I.; Keskar, G.; Majewski, P.W.; Ren, F.; Pfefferle, L.D.; Osuji, C.O. Liquid Crystalline Order and Magnetocrystalline Anisotropy in Magnetically Doped Semiconducting ZnO Nanowires. *ACS Nano* **2011**, *5*, 8357–8364. [CrossRef] [PubMed]

131. Liu, L.H.; Bai, Y.; Wang, F.M.; Liu, N. Fabrication and Characterizes of TiO_2 Nanomaterials Templated by Lyotropic Liquid Crystal. *Adv. Mater. Res.* **2012**, *399–401*, 532–537. [CrossRef]

132. Ren, Z.; Chen, C.; Hu, R.; Mai, K.; Qian, G.; Wang, Z. Two-Step Self-Assembly and Lyotropic Liquid Crystal Behavior of TiO_2 Nanorods. *J. Nanomater.* **2012**, *2012*, 180989. [CrossRef]

133. Li, L.; Walda, J.; Manna, L.; Alivisatos, A.P. Semiconductor Nanorod Liquid Crystals. *Nano Lett.* **2002**, *2*, 557–560. [CrossRef]

134. Li, L.-S.; Alivisatos, A.P. Semiconductor Nanorod Liquid Crystals and Their Assembly on a Substrate. *Adv. Mater.* **2003**, *15*, 408–411. [CrossRef]

135. Kim, F.; Kwan, S.; Akana, J.; Yang, P. Langmuir–Blodgett Nanorod Assembly. *J. Am. Chem. Soc.* **2001**, *123*, 4360–4361. [CrossRef] [PubMed]

136. Thorkelsson, K.; Bai, P.; Xu, T. Self-assembly and applications of anisotropic nanomaterials: A review. *Nano Today* **2015**, *10*, 48–66. [CrossRef]

137. Georgakilas, V.; Otyepka, M.; Bourlinos, A.B.; Chandra, V.; Kim, N.; Kemp, K.C.; Hobza, P.; Zboril, R.; Kim, K.S. Functionalization of Graphene: Covalent and Non-Covalent Approaches, Derivatives and Applications. *Chem. Rev.* **2012**, *112*, 6156–6214. [CrossRef] [PubMed]

138. Rezapour, M.R.; Myung, C.W.; Yun, J.; Ghassami, A.; Li, N.; Yu, S.U.; Hajibabaei, A.; Park, Y.; Kim, K.S. Graphene and Graphene Analogs toward Optical, Electronic, Spintronic, Green-Chemical, Energy-Material, Sensing, and Medical Applications. *ACS Appl. Mater. Interfaces* **2017**, *9*, 24393–24406. [CrossRef] [PubMed]

139. Huang, X.; Yin, Z.; Wu, S.; Qi, X.; He, Q.; Zhang, Q.; Yan, Q.; Boey, F.; Zhang, H. Graphene-Based Materials: Synthesis, Characterization, Properties, and Applications. *Small* **2011**, *7*, 1876–1902. [CrossRef] [PubMed]

140. Singh, V.; Joung, D.; Zhai, L.; Das, S.; Khondaker, S.I.; Seal, S. Graphene based materials: Past, present and future. *Prog. Mater. Sci.* **2011**, *56*, 1178–1271. [CrossRef]
141. Huang, X.; Qi, X.; Boey, F.; Zhang, H. Graphene-based composites. *Chem. Soc. Rev.* **2012**, *41*, 666–686. [CrossRef] [PubMed]
142. Brimicombe, P.D.; Gleeson, H.F. Private communication, 2008.
143. Behabtu, N.; Lomeda, J.R.; Green, M.J.; Higginbotham, A.L.; Sinitskii, A.; Kosynkin, D.V.; Tsentalovich, D.; Parra-Vasquez, A.N.G.; Schmidt, J.; Kesselman, E.; et al. Spontaneous high-concentration dispersions and liquid crystals of graphene. *Nat. Nanotechnol.* **2010**, *5*, 406–411. [CrossRef] [PubMed]
144. Kim, J.E.; Han, T.H.; Lee, S.H.; Kim, J.Y.; Ahn, C.W.; Yun, J.M.; Kim, S.O. Graphene Oxide Liquid Crystals. *Angew. Chem. Int. Ed.* **2011**, *50*, 3043–3047. [CrossRef] [PubMed]
145. Xu, Z.; Gao, C. Graphene chiral liquid crystals and macroscopic assembled fibres. *Nat. Commun.* **2011**, *2*, 571. [CrossRef] [PubMed]
146. Al-Zangana, S.; Iliut, M.; Turner, M.; Vijayaraghavan, A.; Dierking, I. Confinement effects on lyotropic nematic liquid crystal phases of graphene oxide dispersions. *2D Mater.* **2017**, *4*, 41004. [CrossRef]
147. Dan, B.; Behabtu, N.; Martinez, A.; Evans, J.S.; Kosynkin, D.V.; Tour, J.M.; Pasquali, M.; Smalyukh, I.I. Liquid crystals of aqueous, giant graphene oxide flakes. *Soft Matter* **2011**, *7*, 11154–11159. [CrossRef]
148. Jalili, R.; Aboutalebi, S.H.; Esrafilzadeh, D.; Shepherd, R.L.; Chen, J.; Aminorroaya-Yamini, S.; Konstantinov, K.; Minett, A.I.; Razal, J.M.; Wallace, G.G. Scalable One-Step Wet-Spinning of Graphene Fibers and Yarns from Liquid Crystalline Dispersions of Graphene Oxide: Towards Multifunctional Textiles. *Adv. Funct. Mater.* **2013**, *23*, 5345–5354. [CrossRef]
149. Al-Zangana, S.; Iliut, M.; Boran, G.; Turner, M.; Vijayaraghavan, A.; Dierking, I. Dielectric spectroscopy of isotropic liquids and liquid crystal phases with dispersed graphene oxide. *Sci. Rep.* **2016**, *6*, 31885. [CrossRef] [PubMed]
150. Shen, T.-Z.; Hong, S.-H.; Song, J.-K. Electro-optical switching of graphene oxide liquid crystals with an extremely large Kerr coefficient. *Nat. Mater.* **2014**, *13*, 394–399. [CrossRef] [PubMed]
151. Kim, J.Y.; Kim, S.O. Liquid crystals: Electric fields line up graphene oxide. *Nat. Mater.* **2014**, *13*, 325–326. [CrossRef] [PubMed]
152. Zamora-Ledezma, C.; Puech, N.; Zakri, C.; Grelet, E.; Moulton, S.E.; Wallace, G.G.; Gambhir, S.; Blanc, C.; Anglaret, E.; Poulin, P. Liquid Crystallinity and Dimensions of Surfactant-Stabilized Sheets of Reduced Graphene Oxide. *J. Phys. Chem. Lett.* **2012**, *3*, 2425–2430. [CrossRef] [PubMed]
153. Lagerwall, J.P.F.; Scalia, G. *Liquid Crystals with Nano and Microparticles*; World Scientific: Singapore, 2017.
154. Narayan, R.; Kim, J.E.; Kim, J.Y.; Lee, K.E.; Kim, S.O. Graphene Oxide Liquid Crystals: Discovery, Evolution and Applications. *Adv. Mater.* **2016**, *28*, 3045–3068. [CrossRef] [PubMed]
155. Lekkerkerker, H.N.W.; Vroege, G.J. Liquid crystal phase transitions in suspensions of mineral colloids: New life from old roots. *Philos. Trans. A Math. Phys. Eng. Sci.* **2013**, *371*, 20120263. [CrossRef] [PubMed]

nanomaterials

MDPI

Article

Phase Transition-Driven Nanoparticle Assembly in Liquid Crystal Droplets

Charles N. Melton [1], Sheida T. Riahinasab [1], Amir Keshavarz [2], Benjamin J. Stokes [2] and Linda S. Hirst [1,*]

[1] Department of Physics, School of Natural Sciences, University of California, 5200 North Lake Rd., Merced, CA 95343, USA; cmelton@ucmerced.edu (C.N.M.); triahinasab@ucmerced.edu (S.T.R.)
[2] Chemistry and Chemical Biology Unit, School of Natural Sciences, University of California, 5200 North Lake Rd., Merced, CA 95343, USA; akeshavarz@ucmerced.edu (A.K.); bstokes2@ucmerced.edu (B.J.S.)
*   Correspondence: lhirst@ucmerced.edu

Received: 8 February 2018; Accepted: 1 March 2018; Published: 7 March 2018

Abstract: When nanoparticle self-assembly takes place in an anisotropic liquid crystal environment, fascinating new effects can arise. The presence of elastic anisotropy and topological defects can direct spatial organization. An important goal in nanoscience is to direct the assembly of nanoparticles over large length scales to produce macroscopic composite materials; however, limitations on spatial ordering exist due to the inherent disorder of fluid-based methods. In this paper we demonstrate the formation of quantum dot clusters and spherical capsules suspended within spherical liquid crystal droplets as a method to position nanoparticle clusters at defined locations. Our experiments demonstrate that particle sorting at the isotropic–nematic phase front can dominate over topological defect-based assembly. Notably, we find that assembly at the nematic phase front can force nanoparticle clustering at energetically unfavorable locations in the droplets to form stable hollow capsules and fractal clusters at the droplet centers.

Keywords: nematic liquid crystal; quantum dot; nanoparticle; self-assembly; phase transition

1. Introduction

Topological defects in nematic liquid crystals (LCs) are known to drive the self-assembly of included colloidal particles through elastic interactions with the medium [1–3]. Colloidal particles can be located and arranged in two- and three-dimensional packings by the action of defect lines [1] to produce custom lattice-like structures. In general, several types of self-assembled structures have been constructed in different media such as linear chains [4], clusters [5], and structured arrays [6,7]. These self-assembled structures can be driven by a variety of forces, such as kinetics of the particles themselves [8]. Two-dimensional (2D) structures composed of colloidal particles have also been formed on the surface of liquid crystal droplets suspended in water and localized at topological defects on LC droplet surfaces [9–11]. This has been successfully achieved using both nano- and micro-particles [12,13] as well as biological molecules [14], to achieve unique structures.

The assembly of very small particles (~10 nm in diameter or less) presents more of a challenge as the particles become subject to strong Brownian fluctuations where the size of the particles approaches that of the solvent molecules. If the energy scale of these thermal fluctuations (~kT) also becomes comparable to the free energy cost of inserting a particle into the anisotropic liquid crystal medium, spontaneous assembly mediated by the Frank elastic constants [3] can occur. Dispersed particles in the liquid crystal are able to explore the anisotropic fluid thermally and assemble at free energy minima by clustering together and/or locating in topological defect cores.

There have been several recent attempts to use liquid crystal defects to assemble and cluster nanoparticles of several types, including semiconducting (e.g., quantum dots), metallic (e.g., gold, silver), and magnetic particles [15]. Such assemblies may exhibit collective electronic, photonic, or magnetic properties not seen in isolated nanoparticles [16,17]. Typically, particle clustering experiments result in the formation of isolated aggregates with no internal ordering or position control, although the deliberate seeding of defect points [18] or lines [19] can provide an organizing mechanism. An alternative approach to forming nanoparticle assemblies uses monodisperse particles and carefully designed ligands [20]. This work has resulted in fascinating 2D arrangements of particles; however, the technique has not been expanded into large-scale structures in the third dimension.

In this work we were interested in finding a way to reliably direct the spatial organization of nanoparticle clusters and other assemblies. For example, can we make a regular array of small nanoparticle clusters, each of a well-defined size? Our approach is to use liquid crystal droplets in the nematic phase to control the positioning and size of these clusters. Micron-scale droplets containing clusters should be easy to manipulate by external methods (including, optical trapping and surface patterning).

The nematic liquid crystal phase is an anisotropic fluid characterized by orientational order defined locally by the director [21]. Stable topological defects in liquid crystals occur where there is orientational frustration, for example, in bulk, at the center of a spherical nematic droplet or at the poles of a sphere coated with a thin film of smectic liquid crystal [22]. When considering a thin film of nematic liquid crystal, four +1/2 defects occur at locations that form a tetrahedron through a sphere [23]. One way to reliably control the location of these topological defects in liquid crystals is to control the geometry of the material and anchoring conditions at its interfaces. Take for example, a spherical liquid crystal droplet. Homeotropic anchoring conditions (whereby molecules are oriented perpendicular to the interface) will lead to a radial droplet structure, with a hedgehog defect located at the center (Figure 1a,b). In contrast, planar anchoring conditions (whereby molecules lie parallel to the LC/solution interface) will tend to lead to a bipolar structure with two defects located at opposite poles of the droplet (Figure 1c,d).

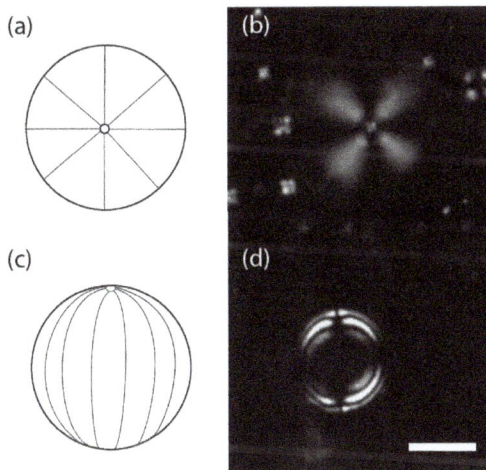

Figure 1. Two common director configurations for a nematic liquid crystal droplet. Radial (single defect in the center): (**a**) schematic; and (**b**) crossed polarizer image of a droplet suspended in aqueous solution, and bipolar (two surface defects); (**c**) schematic; and (**d**) crossed polarizer image of a droplet suspended in aqueous solution. Scale bar = 20 μm.

In previous work, large particles ranging from hundreds of nanometers to several microns in diameter were shown to pin to surface defects in liquid crystal droplets [9]. In these experiments the particle is placed near the defect and subsequently moves to the defect core. We take a different approach, by dispersing nanoparticles into droplets in the isotropic phase, and then subsequently cooling the droplets to the nematic phase. This approach is partially motivated by the inherent difficulties in manipulating individual nanoparticles. It also allows us to begin with a uniform particle distribution and observe cluster formation free from outside manipulation.

To allow spontaneous self-assembly at a defect without the influence of any external force, the particles must be very small and therefore mobile in the liquid crystal phase, taking advantage of Brownian fluctuations to locate at defect points. For this reason, we chose to work with 6-nm quantum dots (QDs) for their bright emission properties, although similar experiments using any nanoparticle type (gold, metal oxide, etc.) would be equivalent.

While self-assembly via topological defect locations is an effective strategy to pursue, recently another mechanism for spatial nanoparticle sorting in liquid crystals was developed [18,20]. Results demonstrated that the moving isotropic to nematic phase front can act as an elastic sorting mechanism for the tiny nanoparticles. Of particular interest is the formation of stable microcapsules or "shells". These structures were formed using quantum dots with mesogenic ligands [24] providing an added degree of control in particle dispersion and cluster stabilization. When closely packed, the mesogenic ligands provide a short-range attractive interaction between nanoparticles.

In this paper we explore these two assembly mechanisms in liquid crystal droplets with different surface anchoring conditions (planar and homeotropic). These two mechanisms we title "equilibrium defect sorting" and "phase transition sorting". By varying the droplet cooling rate through the isotropic to nematic phase transition, we observed different particle distributions in the liquid crystal droplets induced by defect locations and particle assembly at energetically unfavorable locations. In addition, we found that it was also possible to form single nanoparticle microcapsules [24] at the center of liquid crystal droplets.

Understanding the competition between the defect-based assembly and phase-transition-induced assembly is important for controlling the position of nanoparticle clusters over large length-scales without the need for chemical alignment layers or expensive lithography techniques. By investigating assembly with a controlled spherical droplet geometry, we can compare assembly mechanisms more directly, probing the effects of droplet geometry and size, as well as cooling rate, on cluster formation. In addition, we propose that clusters isolated in individual droplets can be close-packed to produce macroscopic assemblies of nanoparticle clusters in two and three dimensions.

2. Materials and Methods

In this work we used quantum dots functionalized with a mesogenic ligand (**8**, Scheme 1). This ligand is an amine-terminated variant of the calamitic side-on attaching liquid crystals investigated by the groups of Dunmur [25] and Vashchenko [26]. It was prepared following the sequence of reactions reported by Quint and coworkers [27], and then exchanged with octadecylamine surface ligands on commercial CdSe core/ZnO shell quantum dots (NN Lab Inc., Fayetteville, AR, USA) following our reported procedure [28]. It was targeted for its ability to stabilize particle clusters via short range non-covalent interactions [19], which would facilitate particle dispersion in the host liquid crystal matrix (4-cyano-4′-pentylbiphenyl, "5CB") and produce a uniform dispersion of modified QDs in the isotropic phase [20]. In addition, above a threshold concentration, the ligand allows the formation of micron-scale capsules [24,27] by a unique phase separation process. Herein, we aimed to form these structures in the more controlled liquid crystal droplet geometry.

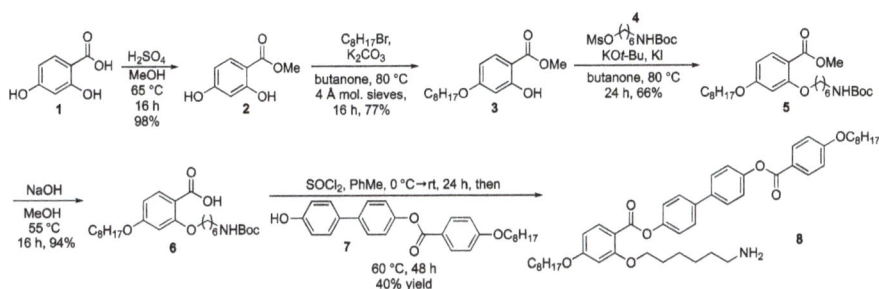

Scheme 1. Sequence of reactions used to prepare the mesogenic ligand.

The degree of ligand exchange under different conditions was quantified using ^{1}H NMR spectroscopy as previously reported [28], revealing a 9:1 surface ratio of **8** to octadecylamine.

The modified particles are uniformly dispersed in 5CB via heat bath sonication at 50 °C for 2 h, verified by fluorescence microscopy. Particle concentrations used in these experiments varied between 0.05 and 0.2 wt %.

Once a uniform particle dispersion in 5CB is achieved, the droplets can be formed. We pipette 3 µL of the QD-5CB composite into 300 µL of either Millipure water or 1.0 wt % polyvinyl alcohol (PVA)/water solution at a temperature of 55 °C. 5CB droplets in water typically exhibit homeotropic boundary conditions resulting in a radial configuration which was verified by cross-polarized microscopy, whereas droplets in PVA/water solution exhibit planar boundary conditions resulting in a bipolar configuration [29], as illustrated in Figure 1c. To follow standard practice for creating nematic droplets, we also dispersed the droplets in a solution of 1 wt % sodium dodecyl sulfate (SDS) to achieve homeotropic boundary conditions, and pure glycerol to produce planar boundary conditions. After adding the QD-5CB composite to the aqueous solution, the system was then tip sonicated using a cell disrupter for approximately one second, until the resulting emulsion appeared cloudy, keeping the system above the nematic–isotropic transition temperature. The rapid motion of the tip sonicator forms droplets of varying sizes in a very small amount of time. Isotropic droplets were then cooled into the nematic phase at two different cooling rates, 1 °C/min and ~200 °C/min, and QD cluster formation and location were observed using a fluorescence microscope.

Fluorescence microscopy was used to image the spatial distribution of QDs in the liquid crystal droplets. In the experiments presented here we used CdSe/ZnS core shell QDs (NN-labs) with an emission wavelength centered at 620 nm. Fluorescence imaging was carried out on an upright Leica DM2500P microscope in reflection mode using a 20× objective. A white-light mercury lamp illumination source with a 515–560 nm band-pass filter was used for QD excitation. Emission was detected using a 580 nm dichroic mirror and a 590 nm long pass filter. The microscope can also be used in transmission mode with a white light source and crossed polarizers to image birefringence. The droplet suspensions were mounted on standard glass slides under a cover slip for observations.

3. Results

3.1. Slow Cooling Experiments

Through our experiments, we utilize two different molecular orientations for the droplets: radial and bipolar. Schematics for these molecular orientations are shown in Figure 1. In our first set of experiments, the liquid crystal droplets with dispersed QDs were cooled at 1 °C/min. Experiments were carried out at two different concentrations of QDs: 0.05 wt % and 0.2 wt %. The lower concentration was specifically chosen to prevent spherical shells and other macroscopic structures from forming via the transition templating process, as we recently reported for the same system in bulk at concentrations above ~0.15 wt % [24], and to obtain a small cluster. Figure 2 shows our results from the slow cooling experiments in which we compared radial and bipolar droplet configurations. Radial droplets resulted in QD clusters localized at the hedgehog defect at the center as shown in Figure 2a. Co-localization was verified using a combination of both fluorescence and cross-polarized microscopy. Cluster sizes varied droplet to droplet, with larger droplets producing larger central clusters. This result is expected, assuming that all droplets begin with a uniform dispersion of QDs and that these dispersed droplets all end up at the central defect after the liquid crystal transitions to the nematic phase.

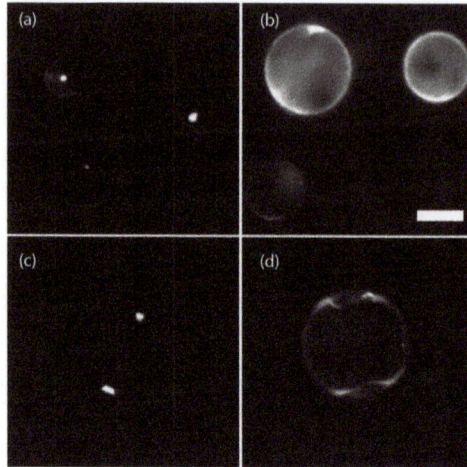

Figure 2. Examples of quantum dot clusters formed via slow cooling of nematic droplets. (**a**) Clusters located at the center of a droplet, with the liquid crystal in a radial configuration; (**b,c**) Two different examples of quantum dot distributions in bipolar droplets at a low particle concentration and (**d**) at a higher particle concentration (For all images, bar = 20 μm)

3.2. Rapid Cooling Experiments

In a second series of experiments we repeated the procedure described above, but with a significantly faster cooling rate of ~200 °C/min. This rapid cooling rate was chosen to match that used in recent experiments where we reported the formation of spherical QD shells by phase transition templating [24]. We first tested the low concentration QD-5CB mixture (0.05 wt %) using homeotropic boundary conditions, and again saw QD clusters forming at the center hedgehog defect. However, when droplets cooled with planar boundary conditions were examined, we observed a surprising result—the QD cluster also formed at the center (Figure 3a)—in contrast to the surface-localized particles exhibited for low cooling rates in Figure 2. Using cross-polarized microscopy, we observed that the cooled droplets in fact had radial defect conformations, not the expected bipolar conformation (Figure 2c for example). This result clearly indicates that within the appropriate parameter range, phase front sorting dominates the assembly process over the slower topological defect assembly process.

We then tested the same fast cooling rate at the higher concentration of QDs in 5CB (0.2 wt %). Cooling these radial droplets produced hollow microshells located at the droplet centers. These microshells are identical to those discussed in our previous publication [24]. However, in this case, we demonstrated that it is possible to form a single hollow shell in the center of a liquid crystal droplet (Figure 4). While the previously reported bulk method for microshell formation is limited by spatial control, this new method provides a mechanism to form individual microshells at specified locations—that is, at the center of LC droplets.

3.3. Cluster Scaling Analysis

To quantify the clustering formation process we carried out a scaling analysis for the phase transition sorting mechanism that produced the central cluster, as shown in Figure 3. Clustering via the two different formation mechanisms produces particle packings that are quite different. Slow particle assembly by cluster-cluster aggregation and subsequent topological defect localization is expected to produce fractal-like packing with a mass-scaling dimension of 1.8 [30,31]. In contrast, the phase front templating method has been shown to produce very dense amorphous particle assemblies, including the micro-shells we demonstrated in Figure 4 [24]. A notable benefit of forming QD clusters in the confined geometry of a droplet is that it gives us the ability to quantify their spatial characteristics—since we know the concentration of particles and the droplet size, we can estimate the mass of quantum dots in each droplet. We can characterize a cluster of nanoparticles by its fractal dimension, D, where the relation between the mass of an object and its size is given as:

$$M = Ar^D \tag{1}$$

where M is the mass of the object, A is a constant, and r represents the radius of the cluster. To calculate the mass-scaling dimension of nanoparticle clusters, we measured cluster size as a function of cluster mass.

In the perfect case, all particles in a droplet would be driven to the central point, and the mass of a specific cluster would simply be obtained as $M = c\frac{4}{3}\pi R^3$, where c is the initial particle concentration before the phase transition (in the isotropic phase) and R is the radius of the droplet. However, we observed that droplet images under fluorescence microscopy indicated some emission in regions other than the central defect. This leads to the conclusion that not all of the nanoparticles were swept up during the isotropic–nematic phase transition and that some remain dispersed in the nematic phase. When estimating the mass scaling dimension of our clusters, we corrected for this effect to obtain more accurate measurements of the cluster mass. To calculate a ratio of the number of nanoparticles in the cluster compared to the number of nanoparticles in the bulk droplet, fluorescence intensity was integrated over the entire droplet using ImageJ and a corrected cluster mass calculated. In addition, background fluorescence away from the droplet was measured and subtracted to account for background noise.

Droplet and cluster diameters were measured using bright field and fluorescence microscopy. To measure cluster diameters, three pixel-wide intensity line profiles were measured from fluorescence images and fitted with a Gaussian profile (Figure 5a). The diameter of the cluster was taken as the full-width half-max of the profile.

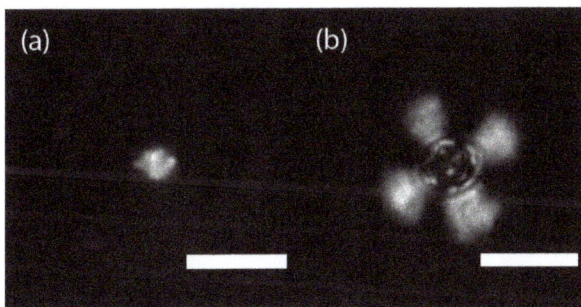

Figure 3. Quantum dot cluster imaged using (**a**) fluorescence microscopy and (**b**) cross-polarized bright field microscopy at the center of a droplet designed to exhibit the bipolar defect configuration. After cooling to the nematic phase, particles were found to be located at a single central point in the radial defect configuration. Scale bars = 20 μm.

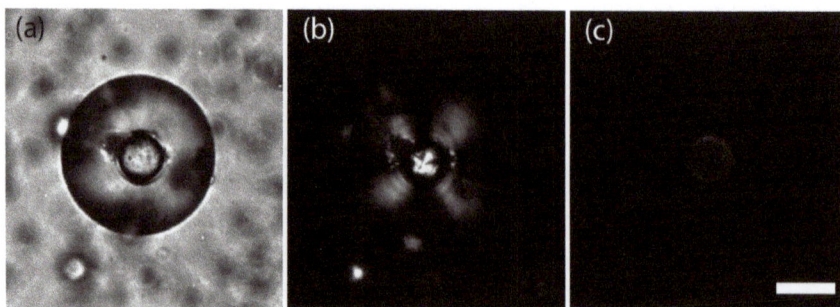

Figure 4. Optical microscopy images of the same droplet with a single quantum dot hollow capsule inside. (**a**) Bright-field image; (**b**) cross-polarized image; and (**c**) fluorescence image, showing quantum dots only. Scale bar = 20 µm.

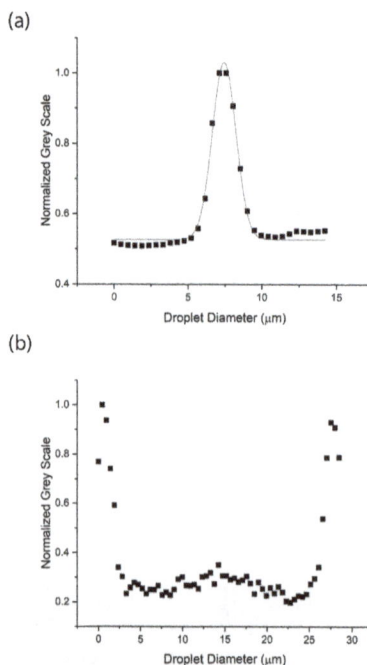

Figure 5. Fluorescence intensity measurements across quantum dot clusters in a droplet as determined using a one-dimensional (1D) line profile across the microscope image. (**a**) Radial droplet example measured across the central cluster (as shown in Figure 1a) showing a peak intensity at the center of a droplet with Gaussian fit; (**b**) Similar data for a bipolar droplet measured across the surface clusters (as shown in Figure 1c) showing two peak intensities at opposite poles of the droplet.

The volume of each droplet was determined, and finally the mass of the droplet was determined using the density of 5CB. Since we know the mass of the 5CB droplet and the QD concentration by wt %, we can calculate the total mass of QDs in each droplet. With the mass and size of the quantum dot clusters, the packing fraction can be obtained. Figure 6 shows cluster mass as a function of cluster radius for an ensemble of different QD clusters formed using the phase front assembly method (rapid cooling, lower concentration). In this plot we assumed a 5% variation on the concentration of quantum dots in isotropic 5CB. The chi-square test resulted in a value of 21. Although not an excellent fit to the data, which exhibits significant scatter, this fit allowed us to quantify the linear trend, observed visually. The slope of this fit represents a scaling dimension of 2.5 ± 0.4, which is relatively dense and consistent with the theoretical value for three-dimensional (3D) ballistic aggregation [30,31].

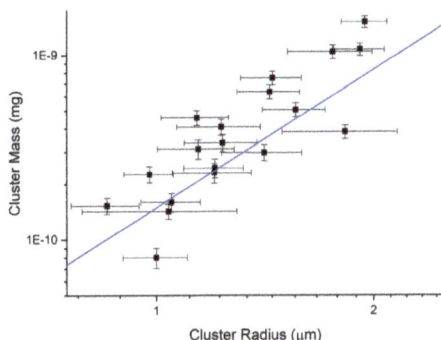

Figure 6. Calculated cluster mass plotted as a function of measured cluster radius. Log–log plot of Equation (1) fit to experimental data. The data collected were cluster mass and cluster size. The straight line fit on a log–log plot indicates a mass scaling dimension of 2.5 ± 0.4.

4. Discussion

Our initial hypothesis was that QD cluster patterning would be achieved through equilibrium defect formation in a liquid crystal droplet in a similar mechanism to that recently reported for micron-scale particles [10–12]. However, we also proposed that rapid cooling might lead to the formation and stabilization of out-of-equilibrium structures such as the recently reported QD spherical shells reported at high cooling rates in a similar system [24]. To investigate these mechanisms, we carried out two sets of experiments, focused on slow cooling and rapid cooling through the isotropic to nematic transition. The droplet geometry was particularly useful in these experiments because the ground-state topological defect configurations are well defined. In addition, the particle concentration within the droplet is easily controlled.

Two Different Mechanisms for Assembly

In this paper we considered two separate mechanisms responsible for the spatial organization of particles in liquid crystal droplets. Overall, the collection of quantum dots occurred at the defect location (i.e., center) of the droplet by a simple free energy argument. There are three major contributions to the free energy for quantum dots inside a liquid crystal droplet: elastic forces, phase transition dynamics (Landau theory (order parameter)), and the actual insertion of quantum dots. This can be expressed as $F = F_{el} + F_{LdG} + F_{QDs}$. The elastic free energy is the classic Frank Free energy which consists of splay, twist, and bend energies; these are the common deformations of a liquid crystal. The Landau de Gennes (LdG) free energy governs the phase transition, and the quantum dot free energy can be considered to represent the particle fraction inside the liquid crystal medium. The particles are swept up due to the changes governed by F_{LdG}, and the location is then governed by the elastic forces that could possibly move the formed cluster to defect locations.

In the first mechanism, "equilibrium defect assembly", the transition particles assemble in the topological defects, producing an 3D aggregate with a fractal-like mass scaling. This is consistent with our previous observations in bulk liquid crystal [20]. Given sufficient time, particles initially not located at defects will eventually migrate there by thermal motion and "fall in" to the aggregate. This process is most likely to occur at very low particle concentrations.

The second mechanism is through "phase transition assembly". In this mechanism, recently reported by our group [20,24], particles preferentially locate in a shrinking isotropic domain within the droplet as it cools through the isotropic to nematic phase transition. The particles are pushed close together as the isotropic phase shrinks to form a cluster.

The first mechanism for nanoparticle sorting in the droplets concerns their tendency to cluster and localize at topological defects in the liquid crystal. Spherical particles of all sizes have been shown to accumulate at defect points in liquid crystals [3]. This is a relatively slow size-dependent process, whereby particles in the nematic phase tend to locate at points of low order to minimize the elastic deformation of the liquid crystal by radially symmetric particles.

The second mechanism is related to the propagation of the isotropic to nematic phase front. In this case, particles preferentially locate in shrinking isotropic domains at the phase transition and are effectively pushed together by the nucleating, growing nematic domains [20,24], entropically repelled by the ordered environment of the nematic phase. This sorting process can occur rapidly and, if the particles are small enough, provides an excellent mechanism to assemble nanoparticle-based structures.

In all of our droplets, we started with a uniform particle distribution in the isotropic liquid crystal phase. Then, following a cooling step, the liquid crystal transitioned into the nematic phase. Our results highlight several possibilities for controlling particle organization.

Using different cooling rates, we observed evidence of competition between the two different assembly mechanisms. Six-nanometer nanoparticle diffusion in the nematic phase can be generally considered as an Brownian random walk, modified by local director orientation. Particles and small clusters will explore the droplet until they locate in a topological defect. We can characterize this motion by timescale, τ_{QD}, the time for a particle to travel a mean squared displacement (MSD) equal to the radius of the droplet.

As nanoparticles and small clusters of particles move randomly, they may encounter other particles or clusters, increasing their size. This will further slow their diffusion rate, increasing τ_{QD}, and thus can be described as cluster-cluster aggregation in an anisotropic environment. Slow cooling the droplet through the phase transition tends to produce many small nematic domains which nucleate and grow. At slow cooling rates this process does not appear to impact the migration of the nanoparticles to topological defects, either at the center of a radial droplet, or on the surface of a bipolar droplet (Figure 1a,c).

Rapid cooling through the phase transition allows the second assembly mechanism to dominate. In this case we typically observe a single-phase front rapidly moving across the droplet. If the phase front transition timescale, τ_{PF} (the time for the phase front to travel the radius of the droplet) is short compared to the QD clustering timescale ($\tau_{PF} < \tau_{QD}$), we expect the phase front particle sorting to dominate over Brownian motion. A simple analysis supports our assumptions. The diffusion constant of a quantum dot in isotropic 5CB was estimated to be 4 $\mu m^2/s$. This was obtained from viscosity measurements from Reference [32] and from using the Einstein–Stokes relation for diffusion, with T above the isotropic–nematic phase transition point. These values yield a linear diffusion length for the QDs of ~5 $\mu m/s$ in a 3D droplet. To quantify the phase front timescale (τ_{PF}), we analyzed a 60-μm diameter droplet and measured the phase front velocity to be 27 $\mu m/s$. These numbers illustrate that we can expect the phase front to move approximately five times faster than the linear diffusion of an average QD in the nematic phase.

We can relate the phase transition assembly process to a non-quasistatic compression. In the case that phase transition assembly dominates, the particles are not able to diffuse away before being swept up by the phase boundary. The particles then follow the direction of the phase front, which depends on the temperature gradient direction in the liquid crystal droplet. If all of the particles are swept up, and the cluster is of sufficient size compared to the size of the droplet, Frank elastic effects are insufficient to move the large cluster, and so the liquid crystal will reorient itself around the cluster to minimize the free energy of the system. This reorientation may not always be the lowest energy state possible, as seen with the surprising example of a cluster stabilized at the center of a bipolar droplet

(Figure 3). This phase front aggregation process could be described as a ballistic diffusion-limited aggregation process as particles are pushed together in the shrinking isotropic domain.

5. Conclusions

In this paper we investigated the spontaneous assembly of nanoparticle clusters in the confined environment of a liquid crystal droplet using mesogen-functionalized quantum dots. Varying surface anchoring conditions at the droplet/water interface allowed us to tune between radial and bipolar topological defect configurations. By tuning particle concentration and cooling rate across the isotropic to nematic phase transition, it was possible to observe the competition between two distinct assembly mechanisms: equilibrium defect assembly and phase transition assembly. We observed two key effects. First, slow cooling allows for ground state defects to template QD cluster formation, while fast cooling allows for the isotropic–nematic phase boundary to template the clusters. Secondly, we noticed that phase transition templating can be used to force non-equilibrium defect configurations, such as a radial director distribution, despite initial bipolar anchoring conditions. This droplet technique also provides a method to control quantum dot micro-shell (and cluster) formation location, opening up the possibility for easy spatial control of micron-scale nanoparticle assemblies.

Acknowledgments: We would like to acknowledge funding from the National Science Foundation CBET-1507551. We also acknowledge Robin Selinger of Kent State University for many thoughtful discussions on liquid crystals and nanoparticles, and Joseph Lopes on his assistance with data analysis techniques.

Author Contributions: C.N.M. conceived and designed the experiments, performed the experiments, and analyzed the data. S.T.R., A.K., and B.J.S. performed the mesogen ligand synthesis and characterization, and functionalized the quantum dots. C.N.M. and L.S.H. wrote the paper.

Conflicts of Interest: The author declares no conflict of interest.

References

1. Poulin, P.; Stark, H.; Lubensky, T.C.; Weitz, D.A. Novel Colloidal Interactions in Anisotropic Fluids. *Science* **1997**, *275*, 1770–1773. [CrossRef] [PubMed]
2. Fleury, J.-B.; Pires, D.; Galerne, Y. Self-Connected 3D Architecture of Microwires. *Phys. Rev. Lett.* **2009**, *103*. [CrossRef] [PubMed]
3. Coursault, D.; Grand, J.; Zappone, B.; Ayeb, H.; Lévi, G.; Félidj, N.; Lacaze, E. Linear Self-Assembly of Nanoparticles Within Liquid Crystal Defect Arrays. *Adv. Mater.* **2012**, *24*, 1461–1465. [CrossRef] [PubMed]
4. Muševič, I.; Škarabot, M.; Tkalec, U.; Ravnik, M.; Žumer, S. Two-Dimensional Nematic Colloidal Crystals Self-Assembled by Topological Defects. *Science* **2006**, *313*, 954–958. [CrossRef] [PubMed]
5. Poulin, P.; Weitz, D.A. Inverted and multiple nematic emulsions. *Phys. Rev. E* **1998**, *57*, 626. [CrossRef]
6. Yada, M.; Yamamoto, J.; Yokoyama, H. Spontaneous formation of regular defect array in water-in-cholesteric liquid crystal emulsions. *Langmuir* **2002**, *18*, 7436–7440. [CrossRef]
7. Urban, J.J.; Talapin, D.V.; Shevchenko, E.V.; Murray, C.B. Self-Assembly of PbTe Quantum Dots into Nanocrystal Superlattices and Glassy Films. *J. Am. Chem. Soc.* **2006**, *128*, 3248–3255. [CrossRef] [PubMed]
8. Bigioni, T.P.; Lin, X.-M.; Nguyen, T.T.; Corwin, E.I.; Witten, T.A.; Jaeger, H.M. Kinetically driven self assembly of highly ordered nanoparticle monolayers. *Nat. Mater.* **2006**, *5*, 265–270. [CrossRef] [PubMed]
9. Rahimi, M.; Roberts, T.F.; Armas-Pérez, J.C.; Wang, X.; Bukusoglu, E.; Abbott, N.L.; de Pablo, J.J. Nanoparticle self-assembly at the interface of liquid crystal droplets. *Proc. Natl. Acad. Sci. USA* **2015**, *112*, 5297–5302. [CrossRef] [PubMed]
10. Whitmer, J.K.; Wang, X.; Mondiot, F.; Miller, D.S.; Abbott, N.L.; de Pablo, J.J. Nematic-Field-Driven Positioning of Particles in Liquid Crystal Droplets. *Phys. Rev. Lett.* **2013**, *111*. [CrossRef] [PubMed]
11. Wang, X.; Miller, D.S.; de Pablo, J.J.; Abbott, N.L. Organized assemblies of colloids formed at the poles of micrometer-sized droplets of liquid crystal. *Soft Matter* **2014**, *10*, 8821–8828. [CrossRef] [PubMed]
12. Blanc, C.; Coursault, D.; Lacaze, E. Ordering nano- and microparticle assemblies with liquid crystals. *Liq. Cryst. Rev.* **2013**, *1*, 83–109. [CrossRef]
13. Hegmann, T.; Qi, H.; Marx, V.M. Nanoparticles in Liquid Crystals: Synthesis, Self-Assembly, Defect Formation and Potential Applications. *J. Inorg. Organomet. Polym.* **2007**, *17*, 483–508. [CrossRef]

14. Wang, X.; Miller, D.S.; Bukusoglu, E.; de Pablo, J.J.; Abbott, N.L. Topological defects in liquid crystals as templates for molecular self-assembly. *Nat. Mater.* **2016**, *15*, 106–112. [CrossRef] [PubMed]

15. Amaral, J.J.; Wan, J.; Rodarte, A.L.; Ferri, C.; Quint, M.T.; Pandolfi, R.J.; Scheibner, M.; Hirst, L.S.; Ghosh, S. Magnetic field induced quantum dot brightening in liquid crystal synergized magnetic and semiconducting nanoparticle composite assemblies. *Soft Matter* **2015**, *11*, 255–260. [CrossRef] [PubMed]

16. Quinten, M.; Leitner, A.; Krenn, J.R.; Aussenegg, F.R. Electromagnetic energy transport via linear chains of silver nanoparticles. *Opt. Lett.* **1998**, *23*, 1331–1333. [CrossRef] [PubMed]

17. Singh, G.; Fisch, M.; Kumar, S. Emissivity and electrooptical properties of semiconducting quantum dots/rods and liquid crystal composites: A review. *Rep. Prog. Phys* **2016**, *79*, 056502. [CrossRef] [PubMed]

18. Rodarte, A.L.; Pandolfi, R.J.; Ghosh, S.; Hirst, L.S. Quantum dot/liquid crystal composite materials: Self-assembly driven by liquid crystal phase transition templating. *J. Mater. Chem. C* **2013**, *1*, 5527. [CrossRef]

19. Guo, Y.; Jiang, M.; Peng, C.; Sun, K.; Yaroshchuk, O.; Lavrentovich, O.; Wei, Q.-H. High-Resolution and High-Throughput Plasmonic Photopatterning of Complex Molecular Orientations in Liquid Crystals. *Adv. Mater.* **2016**, *28*, 2353–2358. [CrossRef] [PubMed]

20. Rodarte, A.L.; Nuno, Z.S.; Cao, B.H.; Pandolfi, R.J.; Quint, M.T.; Ghosh, S.; Hein, J.E.; Hirst, L.S. Tuning Quantum-Dot Organization in Liquid Crystals for Robust Photonic Applications. *Chem. Phys. Chem.* **2014**, *15*, 1413–1421. [CrossRef] [PubMed]

21. Chaikin, P.M.; Lubensky, T.C. *Principles of Condensed Matter Physics*, 1st ed.; Cambridge University Press: Cambridge, UK, 1995; ISBN 0-521-43224-3.

22. Selinger, R.L.B.; Konya, A.; Travesset, A.; Selinger, J.V. Monte Carlo Studies of the XY Model on Two-Dimensional Curved Surfaces. *J. Phys. Chem. B* **2011**, *115*, 13989–13993. [CrossRef] [PubMed]

23. Shin, H.; Bowick, M.J.; Xing, X. Topological Defects in Spherical Nematics. *Phys. Rev. Lett.* **2008**, *101*. [CrossRef] [PubMed]

24. Rodarte, A.L.; Cao, B.H.; Panesar, H.; Pandolfi, R.J.; Quint, M.; Edwards, L.; Ghosh, S.; Hein, J.E.; Hirst, L.S. Self-assembled nanoparticle micro-shells templated by liquid crystal sorting. *Soft Matter* **2015**, *11*, 1701–1707. [CrossRef] [PubMed]

25. Diez, S.; Dunmur, D.A.; Rosario De La Fuente, M.; Karahaliou, P.K.; Mehl, G.; Meyer, T.; Ángel PerÉz Jubindo, M.; Photinos, D.J. Dielectric studies of a laterally-linked siloxane ester dimer. *Liq. Cryst.* **2003**, *30*, 1021–1030. [CrossRef]

26. Prodanov, M.F.; Pogorelova, N.V.; Kryshtal, A.P.; Klymchenko, A.S.; Mely, Y.; Semynozhenko, V.P.; Krivoshey, A.I.; Reznikov, Y.A.; Yarmolenko, S.N.; Goodby, J.W.; et al. Thermodynamically Stable Dispersions of Quantum Dots in a Nematic Liquid Crystal. *Langmuir* **2013**, *29*, 9301–9309. [CrossRef] [PubMed]

27. Quint, M.T.; Sarang, S.; Quint, D.A.; Keshavarz, A.; Stokes, B.J.; Subramaniam, A.B.; Huang, K.C.; Gopinathan, A.; Hirst, L.S.; Ghosh, S. Plasmon-Actuated Nano-Assembled Microshells. *Sci. Rep.* **2017**. [CrossRef] [PubMed]

28. Riahinasab, S.T.; Elbaradei, A.; Keshavarz, A.; Stokes, B.J.; Hirst, L.S. Nanoparticle Microstructures Templated by Liquid Crystal Phase-Transition Dynamics. *Proc. SPIE* **2017**, *10125*, 1012503. [CrossRef]

29. Fernández-Nieves, A.; Link, D.R.; Márquez, M.; Weitz, D.A. Topological Changes in Bipolar Nematic Droplets under Flow. *Phys. Rev. Lett.* **2007**, *98*. [CrossRef] [PubMed]

30. Eggersdorfer, M.L.; Pratsinis, S.E. Agglomerates and aggregates of nanoparticles made in the gas phase. *Adv. Powder Technol.* **2014**, *25*, 71–90. [CrossRef]

31. Botet, R.; Jullien, R. Intrinsic anisotropy of clusters in cluster-cluster aggregation. *J. Phys. A* **1986**, *19*, L907. [CrossRef]

32. Qiao, X.; Zhang, X.; Guo, Y.; Yang, S.; Tian, Y.; Meng, Y. Boundary layer viscosity of CNT-doped liquid crystals: Effects of phase behavior. *Rheol. Acta* **2013**, *52*, 939–947. [CrossRef]

nanomaterials

MDPI

Article

Modifying Thermal Switchability of Liquid Crystalline Nanoparticles by Alkyl Ligands Variation

Jan Grzelak, Maciej Żuk, Martyna Tupikowska and Wiktor Lewandowski *

Faculty of Chemistry, University of Warsaw, 00-927 Warsaw, Poland; jan93nano@gmail.com (J.G.);
maciej.alek.zuk@gmail.com (M.Ż.); m.tupikowska@student.uw.edu.pl (M.T.)
* Correspondence: wlewandowski@chem.uw.edu.pl; Tel.: +48-22-552-6282

Received: 1 February 2018; Accepted: 1 March 2018; Published: 7 March 2018

Abstract: By coating plasmonic nanoparticles (NPs) with thermally responsive liquid crystals (LCs) it is possible to prepare reversibly reconfigurable plasmonic nanomaterials with prospective applications in optoelectronic devices. However, simple and versatile methods to precisely tailor properties of liquid-crystalline nanoparticles (LC NPs) are still required. Here, we report a new method for tuning structural properties of assemblies of nanoparticles grafted with a mixture of promesogenic and alkyl thiols, by varying design of the latter. As a model system, we used Ag and Au nanoparticles that were coated with three-ring promesogenic molecules and dodecanethiol ligand. These LC NPs self-assemble into switchable lamellar (Ag NPs) or tetragonal (Au NPs) aggregates, as determined with small angle X-ray diffraction and transmission electron microscopy. Reconfigurable assemblies of Au NPs with different unit cell symmetry (orthorhombic) are formed if hexadecanethiol and 1H,1H,2H,2H-perfluorodecanethiol were used in the place of dodecanethiol; in the case of Ag NPs the use of 11-hydroxyundecanethiol promotes formation of a lamellar structure as in the reference system, although with substantially broader range of thermal stability (140 vs. 90 °C). Our results underline the importance of alkyl ligand functionalities in determining structural properties of liquid-crystalline nanoparticles, and, more generally, broaden the scope of synthetic tools available for tailoring properties of reversibly reconfigurable plasmonic nanomaterials.

Keywords: reversibly reconfigurable; dynamic self-assembly; nanoparticle assemblies; superlattices; morphing nanoparticles; plasmonic nanoparticles

---

1. Introduction

Nanoparticle (NP) assemblies are formidable materials that can solve a variety of technological problems, such as overcoming speed and bandwidth limitations of modern electronic devices [1,2], surpassing current limits of solar energy harvesting devices [3], retrieving waste heat energy from thermal engines [4], preparing cloaking materials [5], or flat lenses construction [6]. What makes NP assemblies attractive for such wide range of applications are collective phenomena observed when nanocrystals are brought close together. Significantly enhanced electromagnetic field intensities [7], nonlinear optical phenomena [8], plasmonic chirality [9], epsilon near zero properties [10] are just a few examples. Since collective interactions are distance- and direction-sensitive [11,12], a development of methods for the preparation of NP assemblies with tailorable position is highly desired for future optoelectronic technologies.

Self-assembly (SA) is a robust strategy to quickly prepare long-range ordered assemblies of nanocrystals (superlattices) at a low cost [13]. It is most commonly realized by preparing a solution of NPs, followed by the evaporation or destabilization of the solvent, or by the means of gravitational sedimentation. Highly organized structures of NPs can be achieved this way if e.g., template-assisted approach is applied using wrinkled polydimethylsiloxane (PDMS) [14–17], templated PDMS [18] or capillary assembly on a lithographically patterned substrate [19].

Major part of research in the field of nanocrystals self-assembly has been devoted to SA of spherical NPs tethered with hydrocarbon surface ligands (oleylamine, oleic acid, alkyl thiols etc.) which usually behave as quasi-hard spheres, thus assemble into contact area-minimizing configurations (body centered cubic, BCC, face centered cubic, FCC and hexagonal close-packed, HCP) [13]. However, the relatively low diversity of symmetries, static nature and low thermal durability of these assemblies limit their applicability. Thus, strategies for preparing and tuning the response of active [20], non-close packed [21] NP-based materials are required.

Tethering functional molecules to nanoparticle surface has been proposed to guide nanoparticle self-assembly in two-dimensional (2D) and three-dimensional (3D) [22]. A clever choice of surface ligands can soften the particles (e.g., with dendritic [23] or polymer [24] ligands), introduce specific interactions (e.g., with DNA [25], charged [26] and protein [27] ligands) or allow efficient mixing with matrix material [28]; these approaches broaden the scope of available NPs-based structures. Ligand choice is crucial also for preparing active (dynamic, reversibly reconfigurable) nanoparticle assemblies. Molecules, such as DNA [29], poly(N-isopropylacrylamide) [30,31], polystyrene [32], and azobenzene derivatives [33] are often used, endowing NPs with chemical, thermal, electric field and optical sensitivity, respectively. Unfortunately, these dynamic materials suffer from a few drawbacks—they are often limited to close-packed symmetry and usually they require a solvent which hampers observation of the collective phenomena.

There are only a few methods that allow for preparing reversibly reconfigurable NPs assemblies in which nanoparticles are packed densely enough to exhibit collective phenomena; covering nanoparticles with liquid-crystalline (LC) ligands [34–41] is one of those methods. This strategy usually relies on partial exchange of the native ligands that cover nanoparticle with LC compounds, resulting in a binary, LC/alkyl, protecting monolayer (such NPs will be referred to as LC NPs). At low temperatures, this organic coating layer can deform due to LC-ligand bundling, which allows for achieving anisotropic assemblies of nanoparticles with spherical core (Figure 1a). Upon heating, LC-ligands undergo melting transition that leads to isotropization of the coating layer. This deformation is reversible and leads to temperature-dependent reconfigurability of LC-coated NP assemblies. From the practical point of view, it is important to note that the LC-based SA strategy has already proven a feasible approach towards NP assemblies that have unique symmetries [42,43], exhibit structural and functional anisotropy [44,45], as well as exhibiting reconfigurability with short switching times [10]. Notably, several design parameters allowing tailoring properties of LC NPs have been identified. In majority, they relate to the (pro)mesogenic ligands: geometry of the LC-molecule [46–49], its volume [50], LC ligand alkyl spacer length [46,47], as well as density of LC-ligands grafting [51]. Much less effort was put to determining the role of alkyl ligands for which only the influence of length was examined and shown to determine the symmetry of the assemblies (by influencing overall shape of the organic coating layer) [47,52], as well as to influence the thermal response of the material (by influencing spatial rearrangement of LC ligands) [46]. However, to achieve full control over structure and function of the assemblies, it is important to further broaden the design toolbox, especially in the context of tuning symmetry and thermal stability of the assemblies. For practical reasons (time, cost), it would be beneficial if such tuning could be achieved without changing the LC-ligand architecture, e.g., by varying alkyl ligands. Motivated by numerous reports on the properties of monolayers of functionalized alkyl(thiol) compounds [53–57], we assumed that the use of such compounds as ligands could provide new means to control properties of LC NPs assemblies. It is worth to note, that functionalized alkyl ligands were used in LC NPs research, however, mainly with the aim of covalent binding with LC-molecules (direct reaction on nanoparticle surface) [42,58–60].

Here, we demonstrate a new, simple, and robust strategy for tuning the symmetry and thermal response of plasmonic, reversibly reconfigurable nanoparticle assemblies via modification of the alkyl co-ligand design. We systematically assess an impact of the co-ligand architecture on the final symmetry of the superlattices and their thermal response. Importantly, all of the reported samples were prepared using the same LC-ligand, thus the developed strategy is less tedious and time consuming than tailoring properties of LC NPs by synthetizing new LC molecules. Practical applicability of the strategy is highlighted by applying it to plasmonic NPs, in contrast to majority of the previous research done in this field. With the strategy reported here, an easy access to active superlattices with on-demand properties becomes more straightforward.

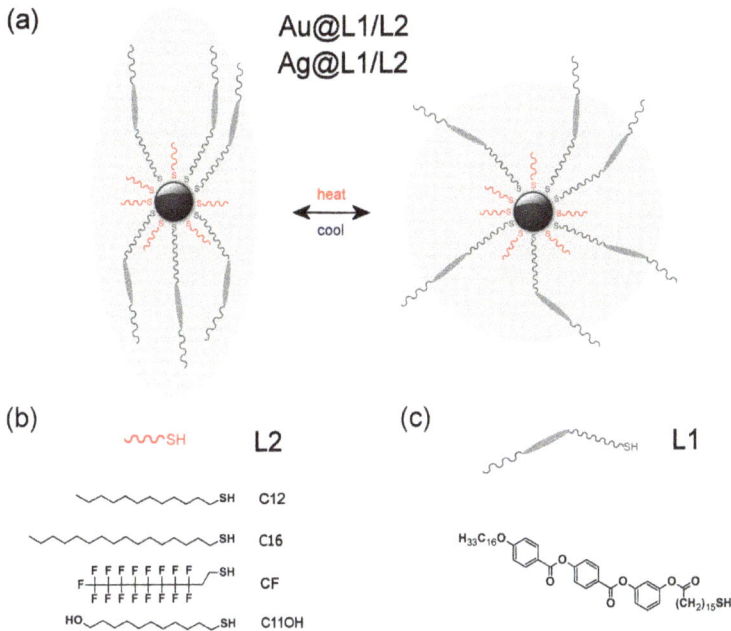

Figure 1. Structure of hybrid silver nanoparticles. (**a**) Schematic model of Au@L1/L2 and Ag@L1/L2 nanoparticles showing thermally driven nanoparticle reshaping resulting from ligand spatial arrangement around nanocrystal core. Limited number of ligands is shown for clarity. Sizes of ligands and nanocrystal core are not to scale. (**b**) Structure of alkyl co-ligands used in the study. (**c**) Structure of the L1 ligand.

2. Results and Discussion

2.1. Designing LC Nanoparticles

The strategy of preparing reversibly reconfigurable superlattices that is reported here is based on covering inorganic nanoparticles' surface with a binary organic shell that comprises liquid-crystalline compounds and alkyl co-ligands. Thus, to prepare the hybrid materials, three aspects of their design had to be considered—nanoparticle core, LC-ligand design, and alkyl co-ligand structure.

As the basis for building liquid crystalline nanomaterials (Figure 1a), we have decided to use small, plasmonic nanoparticles. On one hand, the small size of NPs is desirable, since, according to our previous efforts, only at size range below 6 nm we were able to obtain long-range ordered structures by covering NPs with relatively small (pro)mesogenic ligands [45,52,61]. It should be noted that achieving LC phases with larger nanoparticles is also possible [62], but usually it requires the use of branched (dendritic) ligands. On the other hand, we wanted nanocrystals to exhibit plasmonic properties. Both these conditions (structural and functional) are met by gold and silver nanoparticles (Au and Ag NPs) of few nanometer diameter (size range 3–6 nm). Such NPs do not exhibit as strong plasmonic properties as larger ones [27,63,64], however sub-10 nm Ag NPs are interesting in the context of their prospective applications in surface enhanced Raman spectroscopy [65–67] and for metamaterials preparation [68].

Our hybrid nanomaterials were designed to be covered with a mixture of thiol ligands—a promesogenic molecule (L1, Figure 1c) and one of the four alkyl thiols: dodecanethiol (Au and Ag NPs), hexadecanethiol (Au), 1H,1H,2H,2H-perfluorodecanethiol (Au) and 11-hydroxyundecanethiol (Ag); in general, they will be referred to as L2 ligands (Figure 1b). Nanoparticles bearing L1 and dodecanethiol ligands (Au@L1/C12, Ag@L1/C12) were already described [69,70], and here they served as reference samples for comparison with hybrid nanoparticles equipped with longer (Au@L1/C16) or functionalized (Au@L1/CF, Ag@L1/C11OH) alkyl co-ligands. The new co-ligands were chosen in order to investigate the effect of the length of the alkanethiols (previously measured only for very small, non-plasmonic nanoparticles [46,47,52]), tendency of the fluorinated compounds to segregate from organic phase [57], as well as assess the impact of polar head-group [71,72] for –OH functionalized ligand.

To introduce a mixed thiol monolayer to the surface of nanocrystals, we decided to use a ligand exchange reaction. Specifically, weakly bound amines were substituted with mixtures of alkyl and promesogenic thiols. This approach was chosen since it is more efficient than the direct synthesis of nanocrystals in the presence of the final ligands (large losses of organic compounds). Also, our approach is faster than the previously used protocol [69,70] that required two ligand-exchange steps (first amine to alkyl thiol and then partial exchange of the thiol to (pro)mesogenic molecule).

At this point, it is worth to shortly discuss stability of LC-coated NPs in the view of potential optoelectronic applications. Based on the available data, we can say that NPs coated with LC ligands are more stable than analogous NPs coated exclusively with alkyl ligands [10]. Also, the possibility of cycling LC NPs aggregates between different phases at least few times has been proven, even if it is performed in air [70] (which is known to cause oxidation of Ag NPs surface atoms). Moreover, in another research, it has been shown that it is possible to use UV light irradiation to reconfigure assemblies of NPs coated with photo-sensitive LC-compounds [61]. However, it should be stressed out that detailed research on this topic has not been performed yet.

2.2. Characterization of Au@C12 and Ag@C12 Nanoparticles

Au and Ag NPs were prepared using a two phase method proposed by Wang and Chen [73], yielding nanocrystals that are covered with dodecylamine. The amine is weakly bound to the NP surface, enabling consequent exchange of the native ligands with chosen thiols. At first, we have obtained Ag and Au NPs covered exclusively with dodecanethiol (Au@C12, Ag@C12), they were characterized to determine NPs size and size distribution. It should be stressed out that we did not pursue the characterization of NPs covered with dodecylamine since ligand exchange reaction can have large impact on nanoparticle structure. Small angle X-ray scattering of Au@C12 and Ag@C12 materials suspensions in hexane revealed a characteristic diffractogram (Figure S1a,c). By fitting modelled scattering profile to the experimental curve (Figure S1b,d), the diameter of Au@C12 and Ag@C12 NPs was calculated to be 3.6 ± 0.4 and 5.1 ± 0.3 nm diameter, respectively. To confirm X-ray scattering results, we have used transmission electron microscopy, which revealed well-packed, hexagonal monolayers of NPs with a mean diameter of 3.6 ± 0.4 nm for Au@C12 (Figure 2a) and 5.2 ± 0.4 nm for Ag@C12 (Figure 2c). Direct comparison between results of small angle X-ray scattering (SAXS) and transmission electron microscopy (TEM) investigations (Figure 2b,d) is shown by overlaying histograms of NPs sizes calculated from TEM images with probability density curve derived from SAXS data modelling (the modelled curve is shown in Figure S1b,d). Importantly, both of the techniques indicated low size polydispersity of NPs. We also measured absorption of solutions of Au@C12, Ag@C12 nanoparticles (Figure S2) to confirm plasmonic properties of these NPs.

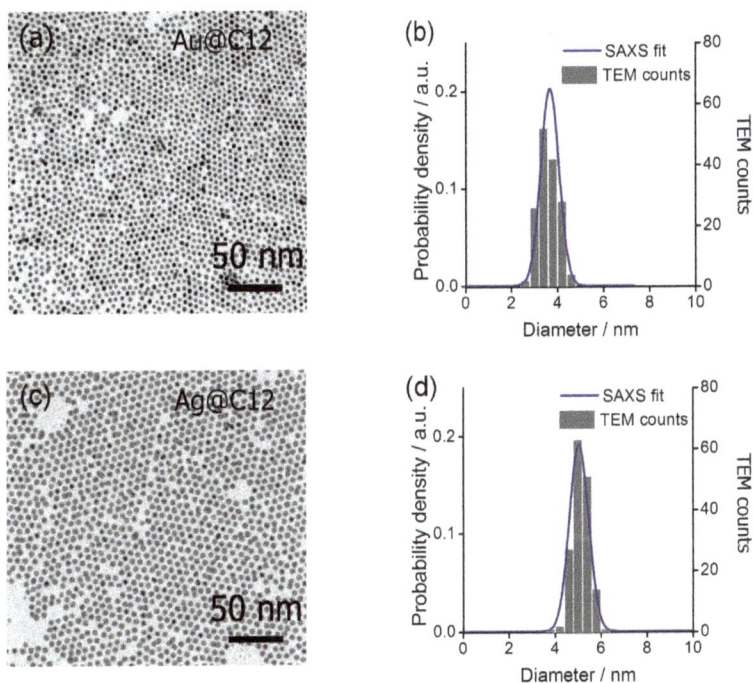

Figure 2. Structural investigation of Au@C12 and Ag@C12 nanoparticles. (**a**) TEM micrograph of Au@C12 monolayer (**b**) Distribution of Au@C12 nanocrystal diameter based on SAXS experiment and TEM analysis. (**c**) TEM micrograph of Ag@C12 monolayer. (**d**) Distribution of Ag@C12 nanocrystal diameter based on SAXS experiment and TEM analysis.

2.3. Preparing and Characterizing Assemblies of Au@L1/C12 and Ag@L1/C12 Nanoparticles

In the next step, we have prepared Au and Ag NPs covered with a mixture of promesogenic (L1) and dodecane-thiols. These samples were designed in analogy to the previously described material and served us as a reference for assessing the role of different alkyl co-ligands. The hybrid NPs were obtained by mixing dodecylamine coated nanocrystals with a 1:1 molar mixture of $C_{12}H_{25}SH$ and L1. The resulting hybrid material was washed several times to remove unbound ligands; the purity of the final product was confirmed using thin layer chromatography. Relative amounts of the ligands introduced to the surface of NPs were measured using ^{1}H NMR technique, as described previously [70], confirming ligand stoichiometry that was close to the reaction mixture (1:1). It is worth to mention that also other techniques, such as X-ray photoelectron spectroscopy (XPS), thermogravimetric analysis (TGA), and elemental analysis (EA) have been shown to give consistent results for estimating composition of the organic coating layer [51]. In future research, it could be also valuable to use surface enhanced Raman spectroscopy (SERS) for this purpose.

Structural investigation of the Ag@L1/C12 NP assemblies was done by small angle X-ray diffraction (SAXRD) measurements. A thick film of the material was prepared on a kapton foil by dropcasting few drops of a concentrated dichloromethane solution of NPs. Then, the film was annealed by keeping the material at 115 °C for 3 min and letting it cool down to 40 °C. Diffractogram obtained for this sample at a low temperature was characteristic for a lamellar structure, it comprised two commensurate signals evidencing layer periodicity, 9.1 nm, and a signal related to mean in-plane interparticle distance, 6.9 nm (Figure 3a). Further confirmation of the phase assignment was achieved by SAXRD measurements of a quasi-monodomain Ag@L1/C12 sample prepared by shearing. The obtained diffractogram (Figure S3) revealed orthogonal azimuthal positions of signals related to layer periodicity and in-plane order, as consistent with the lamellar organization of NPs. Also, transmission electron microscopy investigation of thermally annealed Ag@L1/C12 material (Figure 3d) evidenced lines of nanoparticles with spacing of ca. 8.5 nm that corresponds to inter-layer spacing determined from SAXRD measurements.

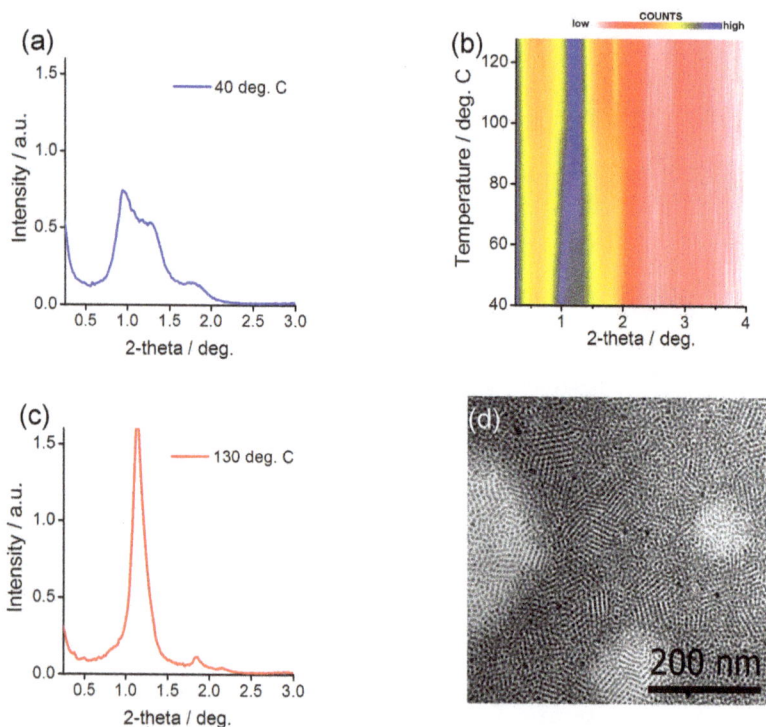

Figure 3. Structural investigation of Ag@L1/C12 nanoparticles. (**a,c**) Small angle X-ray diffraction (SAXRD) profiles for low and high temperature structures (panels a, c, respectively) (**b**) Temperature evolution of SAXRD patterns taken for annealed sample and (**d**) corresponding TEM image.

To test whether the material is thermally reconfigurable, we performed SAXRD experiment while heating the sample. In the range from 30 to 95 °C, a slow monotonic change of the positions of signals was observed, corresponding to the growing distance between NPs within layers (up to 7.4 nm at 90 °C) and a simultaneous decrease of inter-layer spacing (down to 8.2 nm at 90 °C). At ca. 95 °C, a distinct change of the diffractogram was evidenced—above this temperature a set of three narrow signals (Figure 3c) was observed, with temperature independent positions. The best fit to the XRD pattern collected at 130 °C was obtained assuming a FCC structure with unit cell dimension of 13.3 nm.

Structural investigation of the Au@L1/C12 NP assemblies was done in analogy to the above-described experiments. The diffractogram obtained at low temperature (Figure 4a) for heat-annealed sample was the best fit assuming body centered tetragonal symmetry of the aggregate. Unit cell dimensions of Au@L1/C12 NPs assemblies varied with temperature: from 5.1 and 15.8 nm at 30 °C to 5.6 and 14.6 nm at 95 °C. This change corresponds to 10% volume expansion which is reasonable given the difference in temperature and relatively large volume of the organic corona [74]. Also, transmission electron microscopy investigation of thermally annealed Au@L1/C12 material (Figure 4d) evidenced lines of nanoparticles with spacing of ca. 7–8 nm that corresponds well to half the height of the unit cell determined from SAXRD measurements. If heated above 100 °C, thin film of the Au@L1/C12 material undergoes phase transition; above this temperature, only one, relatively broad peak is observed (Figure 4c), evidencing short range ordered structure.

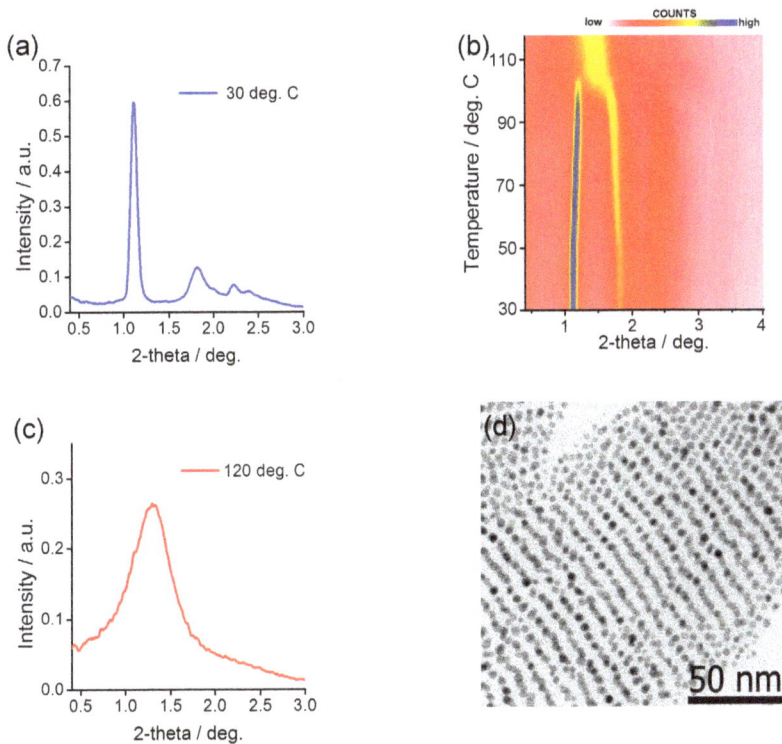

Figure 4. Structural investigation of Au@L1/C12 nanoparticles. (**a,c**) SAXRD profiles for low and high temperature structures (panels a, c, respectively) (**b**) Temperature evolution of SAXRD patterns taken for annealed sample and (**d**) corresponding TEM image.

It is very interesting to compare the behavior of the Au@L1/C12 and Ag@L1/C12 nanoparticles with an analogous sample reported previously. In the case of Ag@L1/C12 nanoparticles in the current study, we observed that heating reorganizes them from long-range ordered lamellar structure to long-range ordered FCC structure, while in previous studies, the long- to short-range order transition was found. Most probably, the origin of the difference lies in a different size distribution of the nanoparticles (6% here vs. 11% in previous research). The lower the size distribution, the better the particles can pack together, promoting the formation of well-ordered structures. This especially applies to high temperature phase in which the organic coating layer is isotropic and preserves spherical shape of the metallic core. In the case of Au@L1/C12 nanoparticles the reverse is truth. Previously, for a similar sample (the difference in one CH_2 moiety [69]), the long- to long-range order switchability was observed, while here, only short range order is observed at high temperatures, which, given the discussion above, might be due to the larger size distribution of NPs used in the current study. However, at low temperatures the same phase was identified evidencing similarity between the samples.

2.4. Preparing and Characterizing Assemblies of Au@L1/C16 Nanoparticles

One of the literature-based strategies to control the structural properties of small LC NPs is the variation of the alkyl co-ligand length. One can argue that the longer the alkyl ligands, the lower are van der Waals forces between nanoparticles, thus the lower is the phase transition temperature. To test this assumption for larger, plasmonic nanoparticles, Au@L1/C16 material has been prepared using the same procedure as for Au@L1/C12 sample. ^{1}H NMR confirmed successful preparation of hybrid material with ca. 1:1 molar ratio of ligands, the same as for Au@L1/C12. Thus, the differences in behavior of Au@L1/C12 and Au@L1/C16 samples can be ascribed to the variation of the alkyl co-ligand length.

At 30 °C, SAXRD measurements of an annealed sample revealed three relatively narrow Bragg peaks that were centered at 8.1, 6.5, and 4.3 nm (Figure 5a). Two approaches have been undertaken to assign the phase—XRD and TEM measurements. First, a sheared sample was prepared that enabled collection of discrete XRD signals (Figure S4). The main signal was positioned along the shearing direction (Figure S5), while for the second, a weak azimuthal splitting was observed giving evidence for 3D long-range order within the sample (Figure S6). Since the signals are quite broad, it would be theoretically possible to fit few different structures of the unit cell, however, by analogy to the analysis of Ag@L1/CF material discussed in the following paragraph, we have chosen base centered orthorhombic structure as the most probable; unit cell dimensions are 6.7, 14.5 and 4.6 nm at 90 °C. Additionally, electron microscopy of heat annealed thin layer of Ag@L1/C16 NPs (Figure 5d) revealed well-ordered rows of NPs that were formed with inter-row distance of ca. 6.5–7.5 nm, corresponding to the half of the *b*-axis unit cell dimension.

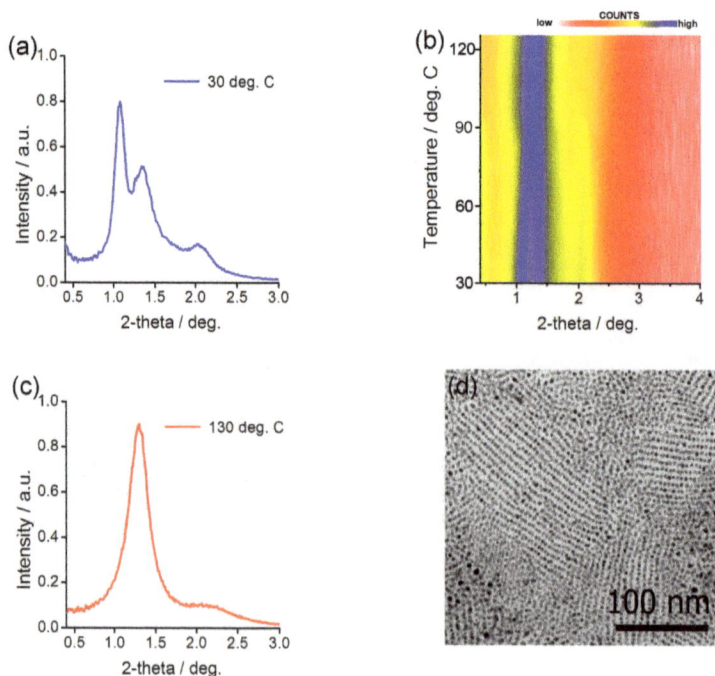

Figure 5. Structural investigation of Au@L1/C16 nanoparticles. (**a,c**) SAXRD profiles for low and high temperature structures (panels a, c, respectively) (**b**) Temperature evolution of SAXRD patterns taken for annealed sample and (**d**) corresponding TEM image.

Besides the difference in symmetry between Au@L1/C12 and Au@L1/C16 NP assemblies, also the stimuli-responsive behavior of the hybrid material has been modified. Namely, XRD experiment with a stepwise heating revealed phase transition at 85 °C, that is 10 °C lower than in the case of Au@L1/C12 NPs. Above this temperature, a pattern (Figure 5c) that could be fitted assuming 12.8 nm FCC unit cell was evidenced. With this experiment we have confirmed that length of alkyl co-ligands can impact symmetry and switchability of plasmonic LC NPs. However, this approach assured a limited success since relatively broad peaks suggest short-range ordered structures were formed.

2.5. Preparing and Characterizing Assemblies of Au@L1/CF Nanoparticles

In a further attempt to vary the self-assembly properties of LC NPs, we decided to use functionalized alkyl-ligands. We focused on commercially available compounds with a chain length that was comparable to dodecanethiol. The thiol we tested was fluorinated decanethiol, namely 1H,1H,2H,2H-perfluorodecanethiol. We have chosen this compound hoping that the ligands would strongly affect self-assembly behavior of the LC NPs due to high tendency of fluorinated compounds to segregate from hydrocarbon matter [57,75].

In analogy to the previous materials, we started structural investigation of the materials from the SAXRD measurements. Diffractogram collected at 30 °C for a thermally annealed structure (Figure 6b) did not allow for us to unequivocally fit any specific symmetry of the unit cell. However, when the sample was heated to 80 °C a clear separation and growth of the intensity of XRD signals was observed (Figure 6a,b) revealing five Bragg signals centered at 6.9, 6.2, 4.8, 3.9, and 3.5 nm (Figure 6a). This pattern can be well-fitted using a base-centered orthorhombic unit cell with the unit cell dimensions of 6.9, 13.8, and 4.8 nm (Figure S7). Measurements performed on the sheared sample further confirmed the assumption (Figure S8), revealing a pattern that is similar to the one observed for Au@L1/C12 material).

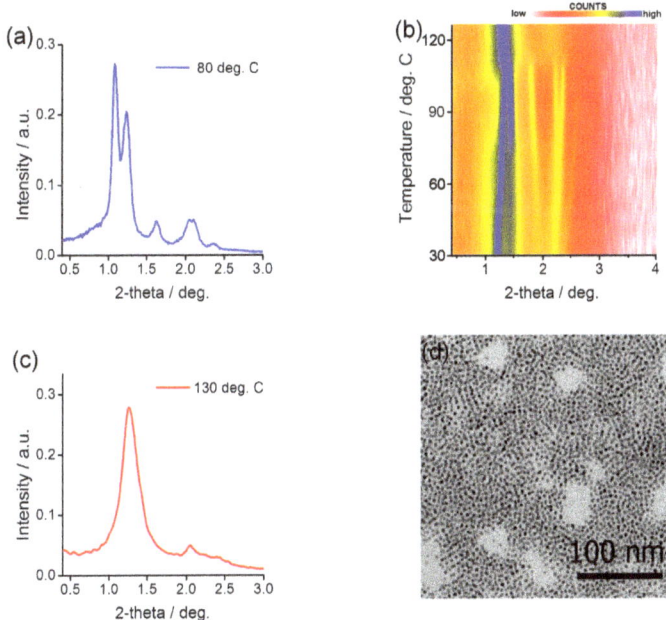

Figure 6. Structural investigation of Ag@L1/CF nanoparticles. (**a,c**) SAXRD profiles for low and high temperature structures (panels a, c, respectively) (**b**) Temperature evolution of SAXRD patterns taken for annealed sample and (**d**) corresponding TEM image.

Further confirmation of the ordered structure formation was achieved with TEM measurements of a heat treated sample (Figure 6d). Weakly developed rows of NPs were evidenced with ca. 7 nm spacing, corresponding to the half of the *b*-axis unit cell dimension. Above 110 °C, a clear change of the diffractogram was observed (Figure 6b,c), evidencing reconfigurability of nanoparticles positions. Phase assignment of this reconfigured structure was performed using a FCC unit cell with 11.9 nm lattice constant.

Results obtained for 1H,1H,2H,2H-perfluorodecanethiol co-ligand attest that it is possible to stabilize anisotropic phase made of LC-grafted NPs by introducing a functional ligand. In this case, also the symmetry of the superlattice was influenced.

2.6. Preparing and Characterizing Assemblies of Ag@L1/C11OH Nanoparticles

Encouraged by results for Au NPs, we decided to test the versatility of our approach to tune properties of LC NPs. Thus, we turned to a different type of nanoparticles (Ag NP) and differently functionalized alkyl co-ligand (11-mercaptoundecanethiol). We assumed that the introduction of polar headgroup could be useful for stabilizing nanoparticle assemblies by dipole-dipole interactions and hydrogen bonding. As for the above discussed samples during the preparation step, we used a 1:1 molar mixture of L1 and C11OH to prepare hybrid NPs via ligand exchange reaction. However, 1H NMR studies of a purified sample revealed that ligand stoichiometry at nanoparticle surface are different from that of parent reaction mixture. Namely, C11OH to L1 thiols molar ratio was estimated as 2:1. This difference can be explained by cooperative ligand exchange process in which attachment of one 11-hydroxyundecanethiol molecule to the surface of a nanoparticle promotes grafting of the same type of ligand. We have performed XRD investigation of as obtained material at low temperatures, which revealed one broad signal, suggesting only short range order of the nanoparticle aggregate. The probable explanation is that there is not enough promesogenic molecules present at the nanoparticle surface to assure efficient deformation of the organic stabilizing layer. Thus, we decided to change stoichiometry of the (ligand exchange) reaction mixture in the search of a sample exhibiting long range order. We succeeded for a mixture comprising 2 molar equivalents of L1 and 1 molar equivalent of C11OH for which 1H NMR studies evidenced 1:1 ligand ratio at NPs surface. In future research it would be also of value to determine the phase separation behavior of functionalized alkyl and LC ligands at the surface of NPs [76]. On one hand, for the functionalized alkyl co-ligands, it seems plausible that enthalpic contribution of phase separation of ligands is larger than for the non-functionalized alkyl co-ligands [77]. On the other hand, an opposing interfacial entropic effect (extra freedom for the longer ligands at the interface with alkyl co-ligands) should also play a role. Thus, for different compositions of the LC/alkyl coating layer various types of ligands distribution at the NPs surface could be obtained (e.g., Janus or patchy); these effects could thus influence the thermal stability and symmetry of the assemblies.

For the latter sample, called Ag@L1/C11OH, SAXRD measurements of thermally annealed material evidenced two relatively narrow signals that were centered at 10.4 and 6.2 nm (Figure 7a,b), which, in analogy to Ag@L1/C12 sample, could be interpreted as inter- and in-plane distances, respectively, for nanoparticles arranged in lamellar structure. An unequivocal proof for the assignment was provided by preparing a quasi-monodomain structure to collect diffractogram with discrete XRD reflections. A pattern comprising signals positioned in directions along and orthogonal to the shearing direction was obtained, characteristic of layered material (Figure S9). Further confirmation of the sample tendency to adopt layered structure was achieved with TEM, which (for heat annealed sample) revealed rows of nanoparticles (Figure 7d) with inter-row distance of ca. 9–10 nm corresponding to XRD derived data and confirming the tendency of the sample to adopt lamellar organization.

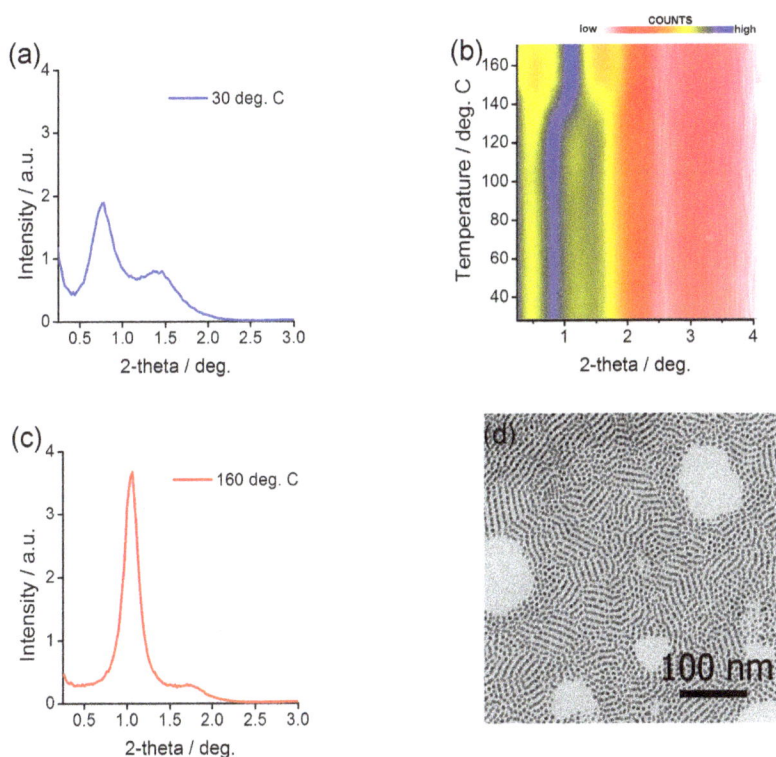

Figure 7. Structural investigation of Ag@L1/C11OH nanoparticles. (**a,c**) SAXRD profiles for low and high temperature structures (panels a, c, respectively) (**b**) Temperature evolution of SAXRD patterns taken for annealed sample and (**d**) corresponding TEM image.

Then, we focused on determining structural reconfigurability of the material by temperature-dependent SAXRD measurements (Figure 6b). The observed XRD signals did not change their positions until temperature 140 °C was reached. This is in clear contrast to the previously described materials for which unit cell dimensions were temperature-dependent, suggesting high structural stability of the Ag@L1/C11OH material. At 140 °C, a clear change in the diffractogram is observed, evidencing phase transition. Notably, the rearrangement takes place at temperature ca. 50 °C higher than for analogous material with non-functionalized alkyl co-ligand (Ag@L1/C12), attesting to a higher stability of the lamellar phase. Above 140 °C, two XRD signals are evidenced. In analogy to the above described samples, we can fit the data with short-range FCC structure that has the unit cell size of 13.5 nm.

Using the SAXRD measured periodicities, we can say that the volume of single Ag@L1/C11OH and Ag@L1/C12 nanoparticles (including the organic coating layer) is very similar, which confirms that similar amounts of LC and alkyl ligands were attached to NPs surface in the ligand exchange reaction. However, larger inter-layer distance (10.4 vs. 9.1 nm, Ag@L1/C11OH vs. Ag@L1/C12 materials, respectively) and shorter in-plane interparticle distance (6.2 vs. 6.9 nm), suggest that these two materials differ in the distribution of the ligands around the nanocrystal core. Namely, the use of functionalized co-ligands assures more efficient separation of ligands in the binary coating layer, thus bundles of the LC ligands are longer, but narrower.

3. Materials and Methods

3.1. Materials

Solvents and substrates were obtained from Sigma-Aldrich (St. Louis, MO 63178, USA). Before use, the solvents were dried over activated molecular sieves for 24 h. Substrates were used without further purification. All of the reactions were carried out in dried glassware with efficient magnetic stirring.

3.2. Preparing LC-NPs Assemblies

To synthetize Ag and Au nanoparticles the modified literature method [67,69] has been used. Shortly, dodecylamine (1.5 g) solution in cyclohexane (50 mL) was stirred for 10 min with 12 mL of aqueous formaldehyde (37%). The organic phase was separated out and washed twice with water (2 × 50 mL). Then, aqueous solution of $AgNO_3$ or $HAuCl_4$ (0.4 g $AgNO_3$ in 20 mL H_2O or 0.08 g $HAuCl_4$ in 20 mL H_2O) was added and left stirring for 40 min. After that, the organic phase was separated and the NPs were precipitated by addition of 100 mL of ethanol. The precipitate was centrifuged, collected, dissolved in a small amount of cyclohexane (10 mL), and the precipitation procedure was repeated again with small amounts of ethanol to enable size fractionation. Usually, 4–5 differently sized fractions of nanoparticles were obtained with narrow size distributions (below 15%).

Then, in the ligand exchange process, LC molecules and corresponding co-ligands were introduced. Shortly, to 15 mg of NPs dissolved in a 6 mL hexane/DCM mixture (1/1, *v/v*), 7 mg of the promesogenic ligand and a stoichiometric amount of the given co-ligand (1.68 mg, 2.15 mg, 3.99 mg, or 1.70 mg of dodecanethiol, hexadecanethiol, 1H,1H,2H,2H-perfluorodecanethiol and 11-hydroxyundecanethiol, respectively) were added. The reaction proceeded at room temperature for 18 h. Then, the solvent was evaporated. The precipitate was dissolved in warm (40 °C) cyclohexane and centrifuged. The supernatant was discarded. Then, the process was repeated until no traces of free ligand molecules remained, as determined by thin-layer chromatography. Nanoparticles were then dissolved in dichloromethane.

The promesogenic ligand has been obtained, according to the previously reported procedure.

3.3. Structural Investigation of LC-NPs Assemblies

1H NMR studies were recorded by using either 200 MHz or 500 MHz NMR Varian Unity Plus. Proton chemical shifts are reported in ppm (δ) relative to the internal standard—tetramethylsilane (TMS δ = 0.00 ppm). For assessing ligand stoichiometry, the nanoparticles were oxidized with I_2 and the reaction mixture after the oxidation process was analyzed.

Transmission electron microscopy (TEM) was performed using Zeiss Libra 120 microscope, with LaB6 cathode, fitted up with OMEGA internal columnar filters and CCD camera. For TEM studies, the solutions of functionalized particles were deposited onto carbon-coated copper grids and then thermally annealed at 120 °C for 3 min and slowly cooled down to 30 °C

The small angle X-ray diffraction (SAXRD) and scattering (SAXS) experiments were realized with the Bruker Nanostar system (CuKα radiation, working in parallel beam geometry formed by cross-coupled Goebel mirrors and 3-pinhole collimation system, area detector VANTEC 2000). The temperature of the sample position was maintained with accuracy of 0.1 °C. Specimens were prepared in thin-walled glass capillaries or as thin films on Kapton tape. For all of the samples, temperature dependent measurements were performed in the same manner—the XRD diffractograms were collected every 5 °C for 300 s with 40 °C/min heating rate between consecutive data collection points.

4. Conclusions

In summary, we introduce a simple and robust approach to tailoring the structural properties of reversibly reconfigurable assemblies of plasmonic nanoparticles. The described nanoparticles were prepared by grafting Ag/Au plasmonic cores with a mixture of promesogenic and straight-chain alkyl thiols. The former were chosen to support nanoparticles self-assembly into long-range ordered structures at low temperatures and to assure temperature-dependent reconfigurability of NP assemblies. The alkyl thiols were however varied—non-functionalized thiols (12- and 16-carbon long) as well as alkyl thiols bearing different functionalities (fluorinated organic chain, hydroxyl moiety) were used. Although relative amounts of ligands at NP surface were kept constant for all of the samples, clear structural differences between the materials were evidenced with SAXRD and TEM measurements. On one hand, with this simple strategy we were able to influence symmetry of the assemblies (lamellar vs. orthorhombic). On the other hand, we were also able to selectively influence thermal stability of assemblies, without changing their symmetry (90 vs. 140 °C melting point for two lamellar systems). It is worth noting that the proposed approach is suitable for plasmonic nanoparticles and allowed for achieving dynamic behavior for all of the tested designs, although further tests are needed to fully assess potential of this approach e.g., in the case of larger nanoparticles or other designs of liquid crystalline ligands. Still, with our proof-of-principle studies, we show that it is worth considering the use of functionalized alkyl ligands to access nanoparticle assemblies with precisely tuned properties. Moreover, since the temperature-driven rearrangement takes place in the neat state (without solvent), we expect that the methodologies that are introduced in this work will promote the design and fabrication of switchable, plasmonic nanomaterials with symmetry and structural stability that is tailored for specific applications in optoelectronics. Future work may also include combining the LC-based approach with template-assisted strategies in order to prepare hierarchically structured, switchable nanomaterials, as well as should focus on testing stability of the LC-based nanomaterials for specific optoelectronic applications.

Supplementary Materials: The following are available online at www.mdpi.com/2079-4991/8/3/147/s1, Figure S1: Structural investigation of Ag@C12 and Au@C12 nanoparticles. (a) SAXS diffractogram of Ag@C12 suspension in hexane. (b) Comparison of modelled (red line) and experimental (circles) 1D SAXS profiles for Ag@C12 material; for modelling the spherical nanoobjects were assumed with diameter 5.1 ± 0.3 nm. (c) SAXS diffractogram of Au@C12 suspension in hexane. (d) Comparison of modelled (red line) and experimental (circles) 1D SAXS profiles for Au@C12 material; for modelling the spherical nanoobjects were assumed with diameter 3.6 ± 0.4 nm, Figure S2: Optical investigation of Ag@C12 and Au@C12 nanoparticles. (a) Absorption spectra of Ag@C12 suspension in hexane. (b) Absorption spectra of Au@C12 suspension in hexane, Figure S3: SAXRD diffractogram of a quasi-monodomain Ag@L1/C12 sample prepared by shearing, Figure S4: SAXRD diffractogram of a quasi-monodomain Au@L1/C16 sample prepared by shearing; measurements at 70 °C, Figure S5: Signal intensity changes along a circle of radius corresponding to (020) signal position in Au@L1/C16 diffractogram shown in Figure S3, Figure S6: Signal intensity changes along a circle of radius corresponding to (110) signal position in Au@L1/C16 diffractogram shown in Figure S3, Figure S7: Signal intensity changes along a circle of radius corresponding to (110) signal position in Au@L1/CF diffractogram shown in Figure 6a in the main text, Figure S8: SAXRD diffractogram of a quasi-monodomain Au@L1/CF sample prepared by shearing; measurements at 80 °C, Figure S9: SAXRD pattern taken for aligned Ag@L1/C11OH sample. Shearing was performed along (10) direction of the structure.

Acknowledgments: This work was supported by the REINFORCE project (agreement No. First TEAM2016-2/15) carried out within the First Team programme of the Foundation for Polish Science co-financed by the European Union under the European Regional Development Fund.

Author Contributions: W.L. conceived and designed the experiments; M.Z., J.G., M.T. and W.L. performed the experiments; W.L. analyzed the data; W.L. wrote the paper.

Conflicts of Interest: The authors declare no conflict of interest.

References

1. Silva, A.; Monticone, F.; Castaldi, G.; Galdi, V.; Alù, A.; Engheta, N. Performing mathematical operations with metamaterials. *Science* **2014**, *343*, 160–163. [CrossRef] [PubMed]

2. Schmidt, D.; Raab, N.; Noyong, M.; Santhanam, V.; Dittmann, R.; Simon, U. Resistive Switching of Sub-10 nm TiO$_2$ Nanoparticle Self-Assembled Monolayers. *Nanomaterials* **2017**, *7*, 370. [CrossRef] [PubMed]

3. Kagan, C.R.; Lifshitz, E.; Sargent, E.H.; Talapin, D.V. Building devices from colloidal quantum dots. *Science* **2016**, *353*, aac5523. [CrossRef] [PubMed]

4. Urban, J.J. Prospects for thermoelectricity in quantum dot hybrid arrays. *Nat. Nanotechnol.* **2015**, *10*, 997–1001. [CrossRef] [PubMed]

5. Ni, X.; Wong, Z.J.; Mrejen, M.; Wang, Y.; Zhang, X. An ultrathin invisibility skin cloak for visible light. *Science* **2015**, *349*, 1310–1314. [CrossRef] [PubMed]

6. Baron, A.; Aradian, A.; Ponsinet, V.; Barois, P. Self-assembled optical metamaterials. *Opt. Laser Technol.* **2016**, *82*, 94–100. [CrossRef]

7. Hamon, C.; Novikov, S.M.; Scarabelli, L.; Solís, D.M.; Altantzis, T.; Bals, S.; Taboada, J.M.; Obelleiro, F.; Liz-Marzán, L.M. Collective Plasmonic Properties in Few-Layer Gold Nanorod Supercrystals. *ACS Photonics* **2015**, *2*, 1482–1488. [CrossRef] [PubMed]

8. Gwo, S.; Wang, C.-Y.; Chen, H.-Y.; Lin, M.-H.; Sun, L.; Li, X.; Chen, W.-L.; Chang, Y.-M.; Ahn, H. Plasmonic Metasurfaces for Nonlinear Optics and Quantitative SERS. *ACS Photonics* **2016**, *3*, 1371–1384. [CrossRef]

9. Wang, Y.; Xu, J.; Wang, Y.; Chen, H. Emerging chirality in nanoscience. *Chem. Soc. Rev.* **2013**, *42*, 2930–2962. [CrossRef] [PubMed]

10. Bagiński, M.; Szmurło, A.; Andruszkiewicz, A.; Wójcik, M.; Lewandowski, W. Dynamic self-assembly of nanoparticles using thermotropic liquid crystals. *Liq. Cryst.* **2016**, *43*, 2391–2409. [CrossRef]

11. Ghosh, S.K.; Pal, T. Interparticle coupling effect on the surface plasmon resonance of gold nanoparticles: From theory to applications. *Chem. Rev.* **2007**, *107*, 4797–4862. [CrossRef] [PubMed]

12. Rožič, B.; Fresnais, J.; Molinaro, C.; Calixte, J.; Umadevi, S.; Lau-Truong, S.; Felidj, N.; Kraus, T.; Charra, F.; Dupuis, V.; et al. Oriented Gold Nanorods and Gold Nanorod Chains within Smectic Liquid Crystal Topological Defects. *ACS Nano* **2017**, *11*, 6728–6738. [CrossRef] [PubMed]

13. Boles, M.A.; Engel, M.; Talapin, D.V. Self-Assembly of Colloidal Nanocrystals: From Intricate Structures to Functional Materials. *Chem. Rev.* **2016**, *116*, 11220–11289. [CrossRef] [PubMed]

14. Steiner, A.M.; Mayer, M.; Seuss, M.; Nikolov, S.; Harris, K.D.; Alexeev, A.; Kuttner, C.; König, T.A.F.; Fery, A. Macroscopic Strain-Induced Transition from Quasi-infinite Gold Nanoparticle Chains to Defined Plasmonic Oligomers. *ACS Nano* **2017**, *11*, 8871–8880. [CrossRef] [PubMed]

15. Mayer, M.; Tebbe, M.; Kuttner, C.; Schnepf, M.J.; König, T.A.F.; Fery, A. Template-assisted colloidal self-assembly of macroscopic magnetic metasurfaces. *Faraday Discuss.* **2016**, *191*, 159–176. [CrossRef] [PubMed]

16. Tebbe, M.; Mayer, M.; Glatz, B.A.; Hanske, C.; Probst, P.T.; Müller, M.B.; Karg, M.; Chanana, M.; König, T.A.F.; Kuttner, C.; et al. Optically anisotropic substrates via wrinkle-assisted convective assembly of gold nanorods on macroscopic areas. *Faraday Discuss.* **2015**, *181*, 243–260. [CrossRef] [PubMed]

17. Hanske, C.; Tebbe, M.; Kuttner, C.; Bieber, V.; Tsukruk, V.V.; Chanana, M.; König, T.A.F.; Fery, A. Strongly coupled plasmonic modes on macroscopic areas via template-assisted colloidal self-assembly. *Nano Lett.* **2014**, *14*, 6863–6871. [CrossRef] [PubMed]

18. Hanske, C.; González-Rubio, G.; Hamon, C.; Formentín, P.; Modin, E.; Chuvilin, A.; Guerrero-Martínez, A.; Marsal, L.F.; Liz-Marzán, L.M. Large-Scale Plasmonic Pyramidal Supercrystals via Templated Self-Assembly of Monodisperse Gold Nanospheres. *J. Phys. Chem. C* **2017**, *121*, 10899–10906. [CrossRef]

19. Flauraud, V.; Mastrangeli, M.; Bernasconi, G.D.; Butet, J.; Alexander, D.T.L.; Shahrabi, E.; Martin, O.J.F.; Brugger, J. Nanoscale topographical control of capillary assembly of nanoparticles. *Nat. Nanotechnol.* **2017**, *12*, 73–80. [CrossRef] [PubMed]

20. Wang, L.; Xu, L.; Kuang, H.; Xu, C.; Kotov, N.A. Dynamic nanoparticle assemblies. *Acc. Chem. Res.* **2012**, *45*, 1916–1926. [CrossRef] [PubMed]

21. Udayabhaskararao, T.; Altantzis, T.; Houben, L.; Coronado-Puchau, M.; Langer, J.; Popovitz-Biro, R.; Liz-Marzán, L.M.; Vuković, L.; Král, P.; Bals, S.; et al. Tunable porous nanoallotropes prepared by post-assembly etching of binary nanoparticle superlattices. *Science* **2017**, *358*, 514–518. [CrossRef] [PubMed]

22. Kuttner, C.; Chanana, M.; Karg, M.; Fery, A. Macromolecular Decoration of Nanoparticles for Guiding Self-Assembly in 2D and 3D. In *Macromolecular Self-Assembly*; Billon, L., Borisov, O., Eds.; John Wiley & Sons, Inc.: Hoboken, NJ, USA, 2016; pp. 159–192, ISBN 9781118887813.

23. Diroll, B.T.; Jishkariani, D.; Cargnello, M.; Murray, C.B.; Donnio, B. Polycatenar Ligand Control of the Synthesis and Self-Assembly of Colloidal Nanocrystals. *J. Am. Chem. Soc.* **2016**, *138*, 10508–10515. [CrossRef] [PubMed]

24. Li, W.; Zhang, P.; Dai, M.; He, J.; Babu, T.; Xu, Y.L.; Deng, R.; Liang, R.; Lu, M.H.; Nie, Z.; et al. Ordering of gold nanorods in confined spaces by directed assembly. *Macromolecules* **2013**, *46*, 2241–2248. [CrossRef]

25. Lin, H.; Lee, S.; Sun, L.; Spellings, M.; Engel, M.; Glotzer, S.C.; Mirkin, C.A. Clathrate colloidal crystals. *Science* **2017**, *355*, 931–935. [CrossRef] [PubMed]

26. Kalsin, A.M.; Fialkowski, M.; Paszewski, M.; Smoukov, S.K.; Bishop, K.J.M.; Grzybowski, B.A. Electrostatic self-assembly of binary nanoparticle crystals with a diamond-like lattice. *Science* **2006**, *312*, 420–424. [CrossRef] [PubMed]

27. Höller, R.P.M.; Dulle, M.; Thomä, S.; Mayer, M.; Steiner, A.M.; Förster, S.; Fery, A.; Kuttner, C.; Chanana, M. Protein-Assisted Assembly of Modular 3D Plasmonic Raspberry-like Core/Satellite Nanoclusters: Correlation of Structure and Optical Properties. *ACS Nano* **2016**, *10*, 5740–5750. [CrossRef] [PubMed]

28. Garbovskiy, Y.; Glushchenko, A. Ferroelectric Nanoparticles in Liquid Crystals: Recent Progress and Current Challenges. *Nanomaterials* **2017**, *7*, 361. [CrossRef] [PubMed]

29. Kim, Y.; Macfarlane, R.J.; Jones, M.R.; Mirkin, C.A. Transmutable nanoparticles with reconfigurable surface ligands. *Science* **2016**, *351*, 579–582. [CrossRef] [PubMed]

30. Li, B.; Smilgies, D.M.; Price, A.D.; Huber, D.L.; Clem, P.G.; Fan, H. Poly(*N*-isopropylacrylamide) surfactant-functionalized responsive silver nanoparticles and superlattices. *ACS Nano* **2014**, *8*, 4799–4804. [CrossRef] [PubMed]

31. Contreras-Cáceres, R.; Pacifico, J.; Pastoriza-Santos, I.; Pérez-Juste, J.; Fernández-Barbero, A.; Liz-Marzán, L.M. Au@pNIPAM thermosensitive nanostructures: Control over shell cross-linking, overall dimensions, and core growth. *Adv. Funct. Mater.* **2009**, *19*, 3070–3076. [CrossRef]

32. Wang, K.; Jin, S.M.; Xu, J.; Liang, R.; Shezad, K.; Xue, Z.; Xie, X.; Lee, E.; Zhu, J. Electric-Field-Assisted Assembly of Polymer-Tethered Gold Nanorods in Cylindrical Nanopores. *ACS Nano* **2016**, *10*, 4954–4960. [CrossRef] [PubMed]

33. Zep, A.; Wojcik, M.M.; Lewandowski, W.; Sitkowska, K.; Prominski, A.; Mieczkowski, J.; Pociecha, D.; Gorecka, E. Phototunable liquid-crystalline phases made of nanoparticles. *Angew. Chem. Int. Ed. Engl.* **2014**, *53*, 13725–13728. [CrossRef] [PubMed]

34. Lewandowski, W.; Wójcik, M.; Górecka, E. Metal nanoparticles with liquid-crystalline ligands: Controlling nanoparticle superlattice structure and properties. *Chemphyschem* **2014**, *15*, 1283–1295. [CrossRef] [PubMed]

35. Bisoyi, H.K.; Kumar, S. Liquid-crystal nanoscience: An emerging avenue of soft self-assembly. *Chem. Soc. Rev.* **2011**, *40*, 306–319. [CrossRef] [PubMed]

36. Stamatoiu, O.; Mirzaei, J.; Feng, X.; Hegmann, T. Nanoparticles in Liquid Crystals and Liquid Crystalline Nanoparticles. In *Topics in Current Chemistry*; Springer: Berlin/Heidelberg, Germany, 2011; Volume 318, pp. 331–393.

37. Qi, H.; Hegmann, T. Liquid crystal–gold nanoparticle composites. *Liq. Cryst. Today* **2011**, *20*, 102–114. [CrossRef]

38. Blanc, C.; Coursault, D.; Lacaze, E. Ordering nano- and microparticles assemblies with liquid crystals. *Liq. Cryst. Rev.* **2013**, *1*, 83–109. [CrossRef]

39. Saliba, S.; Mingotaud, C.; Kahn, M.L.; Marty, J.-D. Liquid crystalline thermotropic and lyotropic nanohybrids. *Nanoscale* **2013**, *5*, 6641–6661. [CrossRef] [PubMed]

40. Nealon, G.L.; Greget, R.; Dominguez, C.; Nagy, Z.T.; Guillon, D.; Gallani, J.-L.; Donnio, B. Liquid-crystalline nanoparticles: Hybrid design and mesophase structures. *Beilstein J. Org. Chem.* **2012**, *8*, 349–370. [CrossRef] [PubMed]

41. Hegmann, T.; Qi, H.; Marx, V.M. Nanoparticles in Liquid Crystals: Synthesis, Self-Assembly, Defect Formation and Potential Applications. *J. Inorg. Organomet. Polym. Mater.* **2007**, *17*, 483–508. [CrossRef]

42. Matsubara, M.; Stevenson, W.; Yabuki, J.; Zeng, X.; Dong, H.; Kojima, K.; Chichibu, S.F.; Tamada, K.; Muramatsu, A.; Ungar, G.; et al. A Low-Symmetry Cubic Mesophase of Dendronized CdS Nanoparticles and Their Structure-Dependent Photoluminescence. *Chem* **2017**, *2*, 860–876. [CrossRef]

43. Cseh, L.; Mang, X.; Zeng, X.; Liu, F.; Mehl, G.H.; Ungar, G.; Siligardi, G. Helically Twisted Chiral Arrays of Gold Nanoparticles Coated with a Cholesterol Mesogen. *J. Am. Chem. Soc.* **2015**, *137*, 12736–12739. [CrossRef] [PubMed]

44. Wolska, J.M.; Pociecha, D.; Mieczkowski, J.; Górecka, E. Control of sample alignment mode for hybrid lamellar systems based on gold nanoparticles. *Chem. Commun.* **2014**, *50*, 7975. [CrossRef] [PubMed]

45. Lewandowski, W.; Constantin, D.; Walicka, K.; Pociecha, D.; Mieczkowski, J.; Górecka, E. Smectic mesophases of functionalized silver and gold nanoparticles with anisotropic plasmonic properties. *Chem. Commun.* **2013**, *49*, 7845–7847. [CrossRef] [PubMed]

46. Lewandowski, W.; Jatczak, K.; Pociecha, D.; Mieczkowski, J. Control of gold nanoparticle superlattice properties via mesogenic ligand architecture. *Langmuir* **2013**, *29*, 3404–3410. [CrossRef] [PubMed]

47. Mang, X.; Zeng, X.; Tang, B.; Liu, F.; Ungar, G.; Zhang, R.; Cseh, L.; Mehl, G.H. Control of anisotropic self-assembly of gold nanoparticles coated with mesogens. *J. Mater. Chem.* **2012**, *22*, 11101–11106. [CrossRef]

48. Kumar, S.; Pal, S.K.; Kumar, P.S.; Lakshminarayanan, V. Novel conducting nanocomposites: Synthesis of triphenylene-covered gold nanoparticles and their insertion into a columnar matrix. *Soft Matter* **2007**, *3*, 896–900. [CrossRef]

49. Demortière, A.; Buathong, S.; Pichon, B.P.; Panissod, P.; Guillon, D.; Bégin-Colin, S.; Donnio, B. Nematic-like organization of magnetic mesogen-hybridized nanoparticles. *Small* **2010**, *6*, 1341–1346. [CrossRef] [PubMed]

50. Kanie, K.; Matsubara, M.; Zeng, X.; Liu, F.; Ungar, G.; Nakamura, H.; Muramatsu, A. Simple cubic packing of gold nanoparticles through rational design of their dendrimeric corona. *J. Am. Chem. Soc.* **2012**, *134*, 808–811. [CrossRef] [PubMed]

51. Wójcik, M.M.; Olesińska, M.; Sawczyk, M.; Mieczkowski, J.; Górecka, E. Controlling the Spatial Organization of Liquid Crystalline Nanoparticles by Composition of the Organic Grafting Layer. *Chemistry* **2015**, *21*, 10082–10088. [CrossRef] [PubMed]

52. Wojcik, M.; Lewandowski, W.; Matraszek, J.; Mieczkowski, J.; Borysiuk, J.; Pociecha, D.; Gorecka, E. Liquid-crystalline phases made of gold nanoparticles. *Angew. Chem. Int. Ed. Engl.* **2009**, *48*, 5167–5169. [CrossRef] [PubMed]

53. Houston, J.E.; Kim, H.I. Adhesion, friction, and mechanical properties of functionalized alkanethiol self-assembled monolayers. *Acc. Chem. Res.* **2002**, *35*, 547–553. [CrossRef] [PubMed]

54. Esplandiú, M.J.; Hagenström, H.; Kolb, D.M. Functionalized self-assembled alkanethiol monolayers on Au(111) electrodes: 1. Surface structure and electrochemistry. *Langmuir* **2001**, *17*, 828–838. [CrossRef]

55. Vericat, C.; Vela, M.E.; Benitez, G.; Carro, P.; Salvarezza, R.C. Self-assembled monolayers of thiols and dithiols on gold: New challenges for a well-known system. *Chem. Soc. Rev.* **2010**, *39*, 1805–1834. [CrossRef] [PubMed]

56. Techane, S.D.; Gamble, L.J.; Castner, D.G. Multi-technique Characterization of Self-assembled Carboxylic Acid Terminated Alkanethiol Monolayers on Nanoparticle and Flat Gold Surfaces. *J. Phys. Chem. C Nanomater. Interfaces* **2011**, *115*, 9432–9441. [CrossRef] [PubMed]

57. Yong, J.; Chen, F.; Yang, Q.; Huo, J.; Hou, X. Superoleophobic surfaces. *Chem. Soc. Rev.* **2017**, *46*, 4168–4217. [CrossRef] [PubMed]

58. Nguyen, T.T.; Nguyen, T.L.A.; Deschenaux, R. Designing liquid-crystalline gold nanoparticles *via* the olefin cross-metathesis reaction. *J. Porphyr. Phthalocyanines* **2016**, *20*, 1060–1064. [CrossRef]

59. Nguyen, T.T.; Albert, S.; Nguyen, T.L.A.; Deschenaux, R. Liquid-crystalline fullerene-gold nanoparticles. *RSC Adv.* **2015**, *5*, 27224–27228. [CrossRef]

60. Mischler, S.; Guerra, S.; Deschenaux, R. Design of liquid-crystalline gold nanoparticles by click chemistry. *Chem. Commun.* **2012**, *48*, 2183–2185. [CrossRef] [PubMed]

61. Wojcik, M.M.; Gora, M.; Mieczkowski, J.; Romiszewski, J.; Gorecka, E.; Pociecha, D. Temperature-controlled liquid crystalline polymorphism of gold nanoparticles. *Soft Matter* **2011**, *7*, 10561. [CrossRef]

62. Yu, C.H.; Schubert, C.P.J.; Welch, C.; Tang, B.J.; Tamba, M.-G.; Mehl, G.H. Design, synthesis, and characterization of mesogenic amine-capped nematic gold nanoparticles with surface-enhanced plasmonic resonances. *J. Am. Chem. Soc.* **2012**, *134*, 5076–5079. [CrossRef] [PubMed]

63. Henry, A.-I.; Courty, A.; Pileni, M.-P.; Albouy, P.; Israelachvili, J. Tuning of solid phase in supracrystals made of silver nanocrystals. *Nano Lett.* **2008**, *8*, 2000–2005. [CrossRef] [PubMed]

64. Wei, J.; Schaeffer, N.; Pileni, M.-P. Ag Nanocrystals: 1. Effect of Ligands on Plasmonic Properties. *J. Phys. Chem. B* **2014**, *118*, 14070–14075. [CrossRef] [PubMed]

65. Wei, J.; Schaeffer, N.; Albouy, P.-A.; Pileni, M.-P. Surface Plasmon Resonance Properties of Silver Nanocrystals Differing in Size and Coating Agent Ordered in 3D Supracrystals. *Chem. Mater.* **2015**, *27*, 5614–5621. [CrossRef]

66. Chapus, L.; Aubertin, P.; Joiret, S.; Lucas, I.T.; Maisonhaute, E.; Courty, A. Tunable SERS Platforms from Small Nanoparticle 3D Superlattices: A Comparison between Gold, Silver, and Copper. *ChemPhysChem* **2017**, *18*, 3066–3075. [CrossRef] [PubMed]

67. Chen, H.Y.; Lin, M.H.; Wang, C.Y.; Chang, Y.M.; Gwo, S. Large-Scale Hot Spot Engineering for Quantitative SERS at the Single-Molecule Scale. *J. Am. Chem. Soc.* **2015**, *137*, 13698–13705. [CrossRef] [PubMed]

68. Young, K.L.; Ross, M.B.; Blaber, M.G.; Rycenga, M.; Jones, M.R.; Zhang, C.; Senesi, A.J.; Lee, B.; Schatz, G.C.; Mirkin, C.A. Using DNA to Design Plasmonic Metamaterials with Tunable Optical Properties. *Adv. Mater.* **2014**, *26*, 653–659. [CrossRef] [PubMed]

69. Lewandowski, W.; Łojewska, T.; Szustakiewicz, P.; Mieczkowski, J.; Pociecha, D. Reversible switching of structural and plasmonic properties of liquid-crystalline gold nanoparticle assemblies. *Nanoscale* **2016**, *8*, 2656–2663. [CrossRef] [PubMed]

70. Lewandowski, W.; Fruhnert, M.; Mieczkowski, J.; Rockstuhl, C.; Górecka, E. Dynamically self-assembled silver nanoparticles as a thermally tunable metamaterial. *Nat. Commun.* **2015**, *6*, 6590. [CrossRef] [PubMed]

71. Gupta, P.; Ulman, A.; Fanfan, S.; Korniakov, A.; Loos, K. Mixed self-assembled monolayers of alkanethiolates on ultrasmooth gold do not exhibit contact-angle hysteresis. *J. Am. Chem. Soc.* **2005**, *127*, 4–5. [CrossRef] [PubMed]

72. Bain, C.D.; Evall, J.; Whitesides, G.M. Formation of Monolayers by the Coadsorption of Thiols on Gold: Valiation in the Head Group, Tail Group, and Solvent. *J. Am. Chem. Soc.* **1989**, *111*, 7155–7164. [CrossRef]

73. Chen, Y.; Wang, X. Novel phase-transfer preparation of monodisperse silver and gold nanoparticles at room temperature. *Mater. Lett.* **2008**, *62*, 2215–2218. [CrossRef]

74. Yu, Y.; Guillaussier, A.; Voggu, V.R.; Houck, D.W.; Smilgies, D.-M.; Korgel, B.A. Bubble Assemblies of Nanocrystals: Superlattices without a Substrate. *J. Phys. Chem. Lett.* **2017**, 4865–4871. [CrossRef] [PubMed]

75. Elbert, K.C.; Jishkariani, D.; Wu, Y.; Lee, J.D.; Donnio, B.; Murray, C.B. Design, Self-Assembly, and Switchable Wettability in Hydrophobic, Hydrophilic, and Janus Dendritic Ligand–Gold Nanoparticle Hybrid Materials. *Chem. Mater.* **2017**, *29*, 8737–8746. [CrossRef]

76. Ong, Q.; Luo, Z.; Stellacci, F. Characterization of Ligand Shell for Mixed-Ligand Coated Gold Nanoparticles. *Acc. Chem. Res.* **2017**, *50*, 1911–1919. [CrossRef] [PubMed]

77. Liu, X.; Yu, M.; Kim, H.; Mameli, M.; Stellacci, F. Determination of monolayer-protected gold nanoparticle ligand–shell morphology using NMR. *Nat. Commun.* **2012**, *3*, 1182. [CrossRef] [PubMed]

![nanomaterials logo] *nanomaterials*

MDPI

Article

Magnetic Nanoparticle-Assisted Tunable Optical Patterns from Spherical Cholesteric Liquid Crystal Bragg Reflectors

Yali Lin [1,2], Yujie Yang [1,2], Yuwei Shan [1,2], Lingli Gong [1,2], Jingzhi Chen [1], Sensen Li [1,2] and Lujian Chen [1,2,*]

[1]   Department of Electronic Engineering, Xiamen University, Xiamen 361005, China;
     phoebe0327@stu.xmu.edu.cn (Y.L.); 23120171152983@stu.xmu.edu.cn (Y.Y.);
     23120171152948@stu.xmu.edu.cn (Y.S.); 23120141153096@stu.xmu.edu.cn (L.G.);
     22920142203654@stu.xmu.edu.cn (J.C.); sensenli@xmu.edu.cn (S.L.)
[2]   Shenzhen Research Institute of Xiamen University, Shenzhen 518057, China
*   Correspondence: lujianchen@xmu.edu.cn; Tel.: +86-592-258-0141

Received: 30 September 2017; Accepted: 2 November 2017; Published: 8 November 2017

Abstract: Cholesteric liquid crystals (CLCs) exhibit selective Bragg reflections of circularly polarized (CP) light owing to their spontaneous self-assembly abilities into periodic helical structures. Photonic cross-communication patterns could be generated toward potential security applications by spherical cholesteric liquid crystal (CLC) structures. To endow these optical patterns with tunability, we fabricated spherical CLC Bragg reflectors in the shape of microshells by glass-capillary microfluidics. Water-soluble magnetofluid with Fe_3O_4 nanoparticles incorporated in the inner aqueous core of CLC shells is responsible for the non-invasive transportable capability. With the aid of an external magnetic field, the reflection interactions between neighboring microshells and microdroplets were identified by varying the mutual distance in a group of magnetically transportable and unmovable spherical CLC structures. The temperature-dependent optical reflection patterns were investigated in close-packed hexagonal arrangements of seven CLC microdroplets and microshells with inverse helicity handedness. Moreover, we demonstrated that the magnetic field-assisted assembly of microshells array into geometric figures of uppercase English letters "L" and "C" was successfully achieved. We hope that these findings can provide good application prospects for security pattern designs.

Keywords: magnetic nanoparticles; cholesteric liquid crystal; Bragg reflection; microfluidics

1. Introduction

The concurrent existence of order and mobility renders liquid crystals (LCs) a unique class of soft functional materials for advanced photonic applications [1–3]. Fortunately, although cholesteric liquid crystals (CLCs) probably could not attract considerable attentions from the industrial community of liquid crystal (LC) displays due to their disadvantages such as slow response time and high driving voltage, they have still enriched the fundamental knowledge of helical superstructures induced by self-assembly and found innovative (non-display) applications based on selective Bragg reflection of circularly polarized (CP) light [4]. Today, there is a burgeoning interest in the use of LCs with unusual aplanar geometries [5–8]. In particular, spherical CLC microstructures with a radial orientation of the helical axes, such as microdroplets and microshells, were investigated as Bragg resonators to construct omnidirectional tunable microlasers operating in the pronounced whispering gallery (WG) mode and distributed feedback (DFB) mode [9–14].

Benefiting from the rapid evolution of microfluidic technologies, the size-polydispersity problem of spherical LC microstructures was successfully overcome, thus paving an attractive way to fabricate sufficiently monodispersed emulsions with controllable geometrical parameters. Recently, photonic cross-communication, arising from light reflections of different wavelengths and handedness orientations in all directions between well-defined spherical CLC microstructures, has generated dynamically tunable multicolored patterns with a specific spatial distribution and has shown potential for chiroptical all-optical distributor/switch and countless security applications [15–18]. The employ of photoresponsive molecular switches enabled a wide tuning range of the pitch length of CLCs and hence of the highly selective CP reflection wavelength emanating from paired spherical CLC Bragg reflectors [16]. In addition, it was reported that the transition from droplets to shells gave rise to sharp patterns and sustained excellent optical quality even after polymerization [19,20].

The arrangement of spherical CLC Bragg reflectors is critical for generating specific patterns. However, to date, the underlying assembly methods to achieve ordered arrays are quite limited, usually induced by flow, gravity, or evaporation, etc. In general, the distance between two nearby droplets/shells cannot be altered arbitrarily and the obtained close-packed hexagonal superstructures usually consist of numerous droplets/shells. Also, it is difficult to manipulate each individual spherical CLC Bragg reflector and organize relatively small amounts of them into various separated geometric figures independently. Very recently, by means of magnetic manipulation strategies, the noncontact transport of CLC microshells was successfully achieved with microfluidic devices [13,19]. Chen et al. fabricated dye-doped CLC microshells encapsulated with water-dispersible Fe_3O_4 nanoparticles for a magnetically-transportable tunable microlaser [13]. Park et al. also demonstrated the capability of solidified CLC microshells with Fe_3O_4 nanoparticles as new types of location-adjustable sensors for the detection of temperature changes, solvent quality, and humidity [19].

Here, we report the microfluidic fabrication of spherical CLC Bragg reflectors in the shape of microshells for magnetic field-assisted optical patterns via photonic cross-communication. Water-soluble magnetofluid consisting of magnetic Fe_3O_4 nanoparticles was selectively incorporated in the inner aqueous core of two types of CLC shells, responsible for the non-invasive transportable capability. We investigated the reflection interaction between neighboring spherical CLC Bragg reflectors that were identified by varying the distance in a group of microshells encapsulated with and without Fe_3O_4 nanoparticles. With the aid of the non-contact control under a magnetic field, CLC droplets and shells with inverse helicity handedness were closely packed into hexagonal arrays. The temperature-dependent tunability of optical reflection patterns were discussed in detail. Moreover, we successfully achieved the magnetic field-assisted assembly of CLC microshells into the geometric figures of uppercase English letters "L" and "C".

2. Materials and Methods

2.1. Materials

Two types of CLC mixtures with inverse helicity handedness were used in the experiment. Mixture I was prepared by adding 2.22 wt % right-handed (RH) chiral dopant R5011 (HCCH) into 97.78 wt % achiral nematic LC E7 (Xianhua, Yantai, China), resulting in the photonic bandgap (PBG) of the CLC locating in the visible light region with the central wavelength around 635 nm. Mixture II was prepared by adding 27 wt % temperature-responsive left-handed (LH) chiral dopant S811 (Xianhua) into 73 wt % E7. The central wavelength of mixture II was about 690 nm at 27 °C and it underwent a blueshift as the temperature increased. The two CLC mixtures were heated above clear point in an oven and mixed ultrasonically until uniform.

Deionized (DI) water dissolved with 10 wt % polyvinyl alcohol (PVA, molecular weight (MW) = 70,000–80,000, 85% hydrolyzed, Aladdin reagent, Shanghai, China) was used as the aqueous phase to enforce planar degenerate anchoring on both inner and outer boundaries, meaning that the LC molecules are forced to lie tangentially near the interfaces. A small proportion of 5 wt % magnetic fluid EMG605 (Ferrotec) consisting of hydrophilic Fe_3O_4 nanoparticles was then homogeneously mixed with the PVA solution to render the microshells magnetically transportable.

2.2. Fabrication of Cholesteric Liquid Crystal Shells/Droplets

Two kinds of glass capillary microfluidic devices were used to fabricate spherical CLC shells and droplets [21], as shown in Figure 1. Figure 1a was used to fabricate monodisperse microshells as a water-in-oil-in-water (W/O/W) double emulsion. By using the EMG605 and PVA solution as the inner phase, we obtained magnetically transportable microshells (hereinafter, *M*-shells). By using PVA solution as the inner phase, we obtained microshells without magnetic transportability (*S*-shells). In both of these two samples, CLC mixture I served as the middle oil phase and PVA solution as the outer aqueous phase. The device in Figure 1b was used to fabricate microdroplets (*T*-droplets) with CLC mixture II as the inner oil phase and PVA solution as the outer aqueous phase. The samples were collected, selected, and sealed in rectangle glass capillaries for further optical observation.

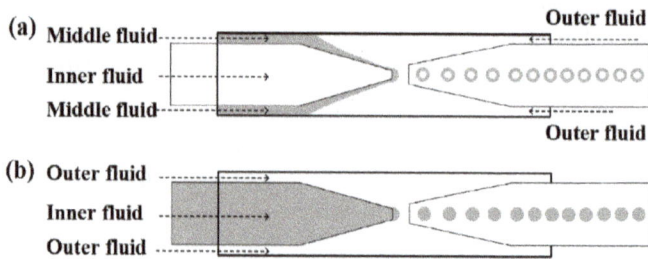

Figure 1. Schematic diagrams of glass capillary microfluidic setups for producing (**a**) water-in-oil-in-water (W/O/W) double emulsion microshells and (**b**) oil-in-water (O/W) microdroplets.

2.3. Optical Characterization

A cross-polarized optical microscope (POM, PM6000, Jiangnan Novel Optics, Nanjing, China) equipped with a charge coupled device (CCD) camera (DCC1645C, Thorlabs, Newton, NJ, USA) was used to measure the size and thickness of microshells and to observe the cross-communication. The numerical aperture (NA) of the objective was 0.25, which means that light with an incident angle smaller than 29° could be collected and measured.

Mixtures I and II were separately filled into planar alignment cells and their reflection spectra were measured at various temperatures. A heating stage (THMS 600, Linkam, Surrey, UK) was used to control the temperature of the samples. A fiber spectrometer (USB4000, Ocean Optics, Shanghai, China) connected to a computer was used to collect the spectra.

3. Experimental Results and Discussions

3.1. Magnetic Movement of CLC Microshells toward Distance-Dependent Reflections

Experimentally, the disclinations in cholesteric droplets and shells were not identified in the reflection mode. So, we suppose that the influence of disclinations on the observation of cross-communication arising from light reflections is weak. In addition, the physical contact between neighboring droplets and shells is avoided by PVA, acting as a surfactant to stabilize the emulsion and preventing the droplets/shells from coalescence and collapse. The phenomenon of the intensity of cross-communication spots between CLC droplets becoming dimmer as their mutual distance increases has been reported previously [16]. One of the main shortcomings in obtaining such intensity-variable optical patterns with cross-communication spots is the randomly-packed structures, since the movement of droplets and their mutual distance cannot be precisely controlled as designed. Nowadays, the separated CLC microshells encapsulated by magnetic nanoparticles are endowed with the ability to be transported, positioned, and gathered together by a magnet. In this experiment, the Fe_3O_4 nanoparticles dispersed in the inner aqueous phase were chemically modified to be hydrophilic and stay in the core owning to the oil-water immiscibility. They were unlikely to immigrate into oil CLC phase and accumulate in the disclinations in the shell [13,19]. This situation is different to that found in lyotropic spherical CLC structures [7,8]. *M*-shells and *S*-shells with the same diameter of ~100 μm and thickness of ~15 μm were chosen to study the dependence of the cross-communication effect on their mutual distance. From a technical point of view, within a short time of the magnetic manipulating process, the *S*-shells without Fe_3O_4 nanoparticles cannot be repositioned by thermal agitation of the outer fluid. Figure 2a–d show the POM images of this process. The upper microshell with a brighter core is the unmovable *S*-shells, while the lower microshell with a darker core is the magnetically transportable *M*-shells with Fe_3O_4 nanoparticles. It was found that the intensity of the reflection spots ascribed to the cross-communication between them became weaker and almost vanished when their distance reached more than 150 μm, as the *M*-shells were moved stepwise away from the *S*-shells. Notably, there are some blurry colored circles in *S*-shells which may possibly be contributed by the internal reflections from the interface between the inner aqueous core and the CLC shell. As for the *M*-shells, the inner core looks much darker because of the light scattering effect inside the CLC shell with the presence of magnetic nanoparticles that are dispersed in the aqueous core.

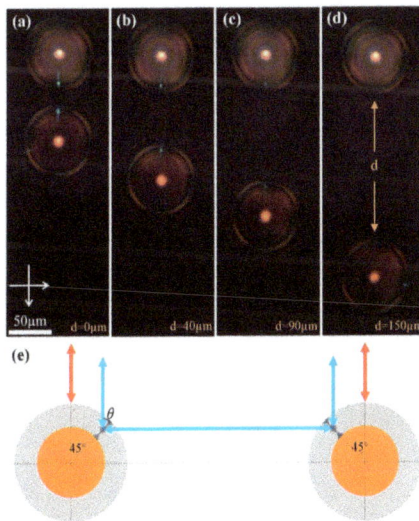

Figure 2. (a–d) Polarized optical microscope (POM) images of the cross-communication phenomenon with distance-dependent intensity. The cross arrow is the mutual position of crossed polarizer. (e) Schematic mechanism of the lateral reflection between two microshells.

Figure 2e is the schematic illustration of the involved mechanism of lateral communication between two microshells with the same pitch and the same helicity handedness. Actually, the density of the aqueous core is lower than that of the LC shell, leading to a potential asymmetric geometry by the interplay between buoyancy and gravity. Meanwhile, a symmetric shell geometry is anticipated to be formed due to the elasticity of the cholesteric helix. Taking all the aforementioned elements into account, we assume that the newly fabricated microshells keep symmetric structures for a long time during the microscopic characterization. The liquid crystal molecules at both inner and outer surfaces of the microshells are planar anchored, resulting in the radial orientation of the helical axes. The incident and reflected lights follow the Bragg condition equation $\lambda = np\cos\theta$, where λ is the wavelength of the incident and reflected lights, n stands for the average refractive index of the CLC, p is the pitch of the CLC, and θ is the incident angle indicated in Figure 2e. When $\theta = 0°$, λ is calculated to be ~635 nm, which means that the central red spot corresponds to the selective reflection of normal incidence. For $\theta = 45°$, the light would reflected to the horizontal direction, enter the contiguous microshells, and reflect again in the vertical direction with the wavelength λ of ~450 nm. The observed red and blue colors corresponding to the central and lateral reflection spots are in accordance with the calculations, respectively.

3.2. Influence of Handedness and Pitch on Tunable Optical Patterns in Close-Packed Hexagonal Arrays with CLC Microdroplets and Microshells by a Magnet

CLCs can spontaneously form into photonic band structures with periodic dielectric helical arrangements. The anisotropic nature of the LC molecules, combined with the continually rotating director *n*, results in the existence of a reflection band for CP light with the same rotation sense as the helix. The co-handed CP reflection is said to be highly sensitive and can only be realized for a small incident angle [22,23]. We chose two CLCs with inverse helicity handedness and different thermal sensitivities to study the tunable optical patterns induced by cross-communication. Since *T*-droplets doped with the LH chiral molecule S811 possess significant thermosensitivity and *M*-shells doped with the RH chiral molecule R5011 are far less sensitive to temperature, we could also easily vary the temperature to examine the reflection of different pitch combinations. As depicted in Figure 3, the wavelength reflection center of mixture II shifts to the blue side from 690 to 570 nm by changing the temperature from 27 to 35 °C. In addition, the inset in Figure 3 confirms the thermo-stable reflection band of mixture I in the temperature range studied.

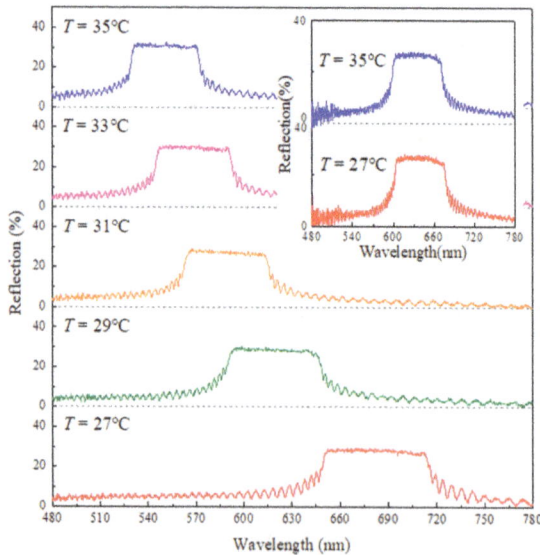

Figure 3. The reflection spectra of thermosensitive cholesteric liquid crystal (CLC) mixture II measured from 27 to 35 °C. The inset shows the thermostability of the reflection band of CLC mixture I in the temperature range studied.

Figure 4a–f show the cross-communications among several *T*-droplets and *M*-shells in close-packed hexagonal arrangements. To make a close-packed hexagonal arrangement, we tried different proportions of these two samples and found that five *M*-shells together with two *T*-droplets or six *M*-shells with one *T*-droplet could form better arrangements. Although the *T*-droplets are unable to be moved by a magnet, they can still be driven by the neighboring *M*-shells. The arrangement in Figure 4a–e is a combination of two *T*-droplets (at the center and upper right, circled in red) and five *M*-shells (2 + 5 combination), while Figure 4f shows a *T*-droplet surrounded by six

M-shells (1 + 6 combination). In the 2 + 5 combination, the cross-communications occur between three combinations of various spherical structures, namely two *M*-shells, two *T*-droplets of same pitch and helicity handedness, and *M*-shells and *T*-droplets with opposite helicity handedness and different pitches.

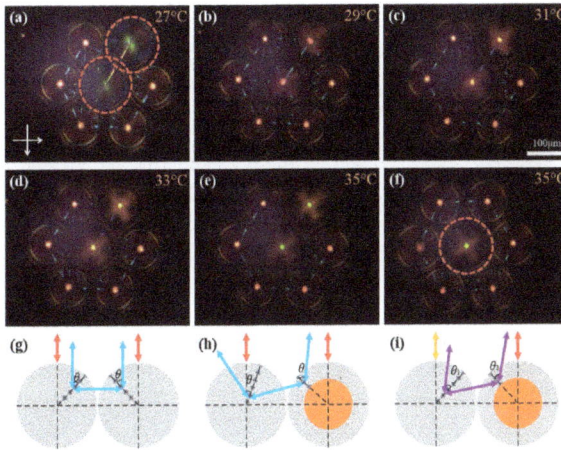

Figure 4. (**a–f**) POM images of the cross-communication between microdroplets and microshells with different pitches and inverse helicity handedness in close-packed hexagonal arrangements. The samples circled in red are *T*-droplets. (**g–i**) Schematic mechanism of the reflection between (**g**) microdroplets, (**h**) microdroplets and microshells with same pitch, and (**i**) microdroplets and microshells with different pitches.

Similar to the microshells discussed above, two CLC microdroplets with the same pitch and helicity handedness could form lateral communication, as shown in Figure 4g. It is worth mentioning that the NA of our objective was 0.25, thus the reflected light could be observed not only in a precisely vertical direction. As a matter of fact, asymmetric reflected path was allowed in a small range of incident and reflected angles [18]. Therefore, the reflections could be established as long as the condition $\lambda = n_1 p_1 \cos\theta_1 = n_2 p_2 \cos\theta_2$ was satisfied, as in the examples shown in Figure 4h,i. In Figure 4a–f, we could clearly identify the reflected spots with different pitches experimentally. As the temperature gradually rose, the pitch of *T*-droplets decreased and the wavelength of all the reflected spots blueshifted until the reflected light reached the invisible ultraviolet region. It was confirmed from Figure 4a,d that the cross-communication between two CLC spherical structures with opposite helicity handedness could still exist, although it was much weaker than that with the same helicity handedness. This agrees well with previous theoretical analysis and experimental results showing that the reflection should involve more complex polarization modes when the incident angle is not equal to zero [17,22,23]. This finding provides a possible way to control the reflected intensity at the same distance by changing the helicity handedness.

3.3. Magnetic Control of Macroscopical Arrays for Secure Authentication

Currently, spherical structures arranged in designed arrays are of particular interest for their promising applications in anti-forgery patterning [20]. Usually, these arrays are formed by depositing particles in pre-defined trenches, holes, or other templates fabricated via mechanical rubbing or photolithography [20,24]. Herein, we proposed a simple way to arrange spherical CLC structures into more complex patterns by taking advantage of their magnetic transportability. As shown in Figure 5a, we sealed a suitable number of *M*-shells in a rectangle glass cell and used a pen with a magnetic tip to manipulate them into uppercase English letters "L" and "C", which were the initial characters of the

Nanomaterials **2017**, *7*, 376

words "Liquid" and "Crystal". These microshells were positioned and arranged into the designed geometric figures, as exhibited in Figure 5b–c. In this manner, we can expect that more intricate patterns could be realized if the magnetic field is controlled precisely. Furthermore, the arrays with designed patterns arranged in this manner are reconfigurable in comparison to the aforementioned template-based methods. If magnetically transportable microshells with different helicity handedness and thermosensitivities are mixed to generate arrays, the spatial distribution of reflection spots with varying colors and intensities would respond to external stimuli, e.g., temperature and light, etc. Such dynamic changes, which are believed to be difficult to forge, can provide good photonic application prospects toward security authentication. Nevertheless, the temporal and mechanical stabilities of photonic cross-communication patterns are crucial for concrete applications, as we discuss above that the thermal agitation of surrounding fluids would possibly disturb the arrangement of shells in the absence of an external magnetic field. It was reported that the full photonic properties of spherical CLCs prepared with a reactive mesogen mixture can still be maintained after the extraction of a nonreactive chiral dopant [25]. Driven mainly by the surface and the interfacial tensions, these solidified CLC microspheres can interconnect with each other and sink into the polymer films they are deposited on after suitable vapor annealing processes [26]. This approach to improve stabilities can be applied in many CLC application fields that were restricted by LCs' unstable fluidic state.

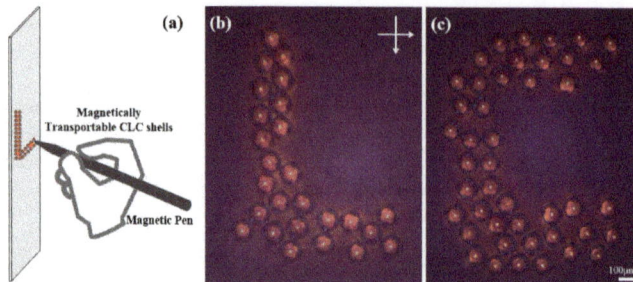

Figure 5. (a) Schematic diagram of controlling microshells by a magnetic pen; (b,c) POM images of the microshells arranged into English letters "L" and "C".

4. Conclusions

In conclusion, we used microfluidic technology to incorporate water-soluble magnetofluid containing Fe_3O_4 nanoparticles into the inner aqueous core of CLC shells, responsible for the non-invasive transportable capability. The reflection interactions between neighboring spherical CLC Bragg reflectors were identified by varying the mutual distance in a group of magnetically transportable and unmovable microshells under a magnetic field. The temperature-dependent tunability of optical reflection patterns was investigated in close-packed hexagonal arrangements of seven CLC droplets and shells with different pitches and inverse helicity handedness. Moreover, we demonstrated that the magnetic field-assisted assembly of microshells arranged into arrays with geometric figures of uppercase English letters "L" and "C" can be successfully achieved for security authentication.

Acknowledgments: This work was financially supported by the National Natural Science Foundation of China (NSFC) (Nos. 61675172 and 61505173), the Natural Science Foundation of Fujian Province, China (No. 2017J01124), Shenzhen Science and Technology Project (No. JCYJ20170306142028457), and the Training Program of Innovation and Entrepreneurship for Undergraduates of Xiamen University (No. 201610384039).

Author Contributions: Lujian Chen and Sensen Li conceived and designed the experiments; Yali Lin, Yuwei Shan, and Lingli Gong performed the experiments; Yujie Yang and Jingzhi Chen analyzed the data; Yali Lin wrote the paper.

Conflicts of Interest: The authors declare no conflict of interest.

References

1. Neill, M.O.; Kelly, S.M. Photoinduced surface alignment for liquid crystal displays. *J. Phys. D Appl. Phys.* **2000**, *33*, 67–84. [CrossRef]
2. Palto, S.P.; Blinov, L.M.; Barnik, M.I.; Lazarev, V.V.; Umanskii, B.A.; Shtykov, N.M. Photonics of liquid-crystal structures: A review. *Crystallogr. Rep.* **2011**, *56*, 622. [CrossRef]
3. Kumar, M.; Kumar, S. Liquid crystals in photovoltaics: A new generation of organic photovoltaics. *Polym. J.* **2017**, *49*, 85–111. [CrossRef]
4. Mitov, M. Cholesteric liquid crystals in living matter. *Soft Matter* **2017**, *13*, 4176–4209. [CrossRef] [PubMed]
5. Urbanski, M.; Reyes, C.G.; Noh, J.; Sharma, A.; Geng, Y.; Subba Rao Jampani, V.; Lagerwall, J.P. Liquid crystals in micron-scale droplets, shells and fibers. *J. Phys. Condens. Matter* **2017**, *29*, 133003. [CrossRef] [PubMed]
6. Muševič, I. Integrated and topological liquid crystal photonics. *Liq. Cryst.* **2013**, *41*, 418–429. [CrossRef]
7. Li, Y.; Jun-Yan Suen, J.; Prince, E.; Larin, E.M.; Klinkova, A.; Therien-Aubin, H.; Zhu, S.; Yang, B.; Helmy, A.S.; Lavrentovich, O.D.; et al. Colloidal cholesteric liquid crystal in spherical confinement. *Nat. Commun.* **2016**, *7*, 12520. [CrossRef] [PubMed]
8. Li, Y.; Prince, E.; Cho, S.; Salari, A.; Mosaddeghian Golestani, Y.; Lavrentovich, O.D.; Kumacheva, E. Periodic assembly of nanoparticle arrays in disclinations of cholesteric liquid crystals. *Proc. Natl. Acad. Sci. USA* **2017**, *114*, 2137–2142. [CrossRef] [PubMed]
9. Humar, M.; Muševič, I. Surfactant sensing based on whispering-gallery-mode lasing in liquid-crystal microdroplets. *Opt. Express* **2011**, *19*, 19836–19844. [CrossRef] [PubMed]
10. Wang, Y.; Li, H.; Zhao, L.; Liu, Y.; Liu, S.; Yang, J. Tunable whispering gallery modes lasing in dye-doped cholesteric liquid crystal microdroplets. *Appl. Phys. Lett.* **2016**, *109*, 231906. [CrossRef]
11. Uchida, Y.; Takanishi, Y.; Yamamoto, J. Controlled Fabrication and Photonic Structure of Cholesteric Liquid Crystalline Shells. *Adv. Mater.* **2013**, *25*, 3234–3237. [CrossRef] [PubMed]
12. Chen, L.J.; Li, Y.N.; Fan, J.; Bisoyi, H.K.; Weitz, D.A.; Li, Q. Photoresponsive Monodisperse Cholesteric Liquid Crystalline Microshells for Tunable Omnidirectional Lasing Enabled by a Visible Light-Driven Chiral Molecular Switch. *Adv. Opt. Mater.* **2014**, *2*, 845–848. [CrossRef]
13. Chen, L.J.; Gong, L.L.; Lin, Y.L.; Jin, X.Y.; Li, H.Y.; Li, S.S.; Che, K.J.; Cai, Z.P.; Yang, C.Y. Microfluidic fabrication of cholesteric liquid crystal core-shell structures toward magnetically transportable microlasers. *Lab. Chip* **2016**, *16*, 1206–1213. [CrossRef] [PubMed]
14. Lin, Y.L.; Gong, L.L.; Che, K.J.; Li, S.S.; Chu, C.X.; Cai, Z.P.; Yang, C.Y.; Chen, L.J. Competitive excitation and osmotic-pressure-mediated control of lasing modes in cholesteric liquid crystal microshells. *Appl. Phys. Lett.* **2017**, *110*, 223301. [CrossRef]
15. Noh, J.; Liang, H.L.; Drevensek-Olenik, I.; Lagerwall, J.P.F. Tuneable multicoloured patterns from photonic cross-communication between cholesteric liquid crystal droplets. *J. Mater. Chem. C* **2014**, *2*, 806–810. [CrossRef]
16. Fan, J.; Li, Y.; Bisoyi, H.K.; Zola, R.S.; Yang, D.K.; Bunning, T.J.; Weitz, D.A.; Li, Q. Light-directing omnidirectional circularly polarized reflection from liquid-crystal droplets. *Angew. Chem.* **2015**, *54*, 2160–2164. [CrossRef] [PubMed]
17. Abetahoff, S.J.; Sukas, S.; Yamaguchi, T.; Hommersom, C.A.; Le Gac, S.; Katsonis, N. Superstructures of chiral nematic microspheres as all-optical switchable distributors of light. *Sci. Rep.* **2015**, *5*, 14183. [CrossRef] [PubMed]
18. Geng, Y.; Noh, J.; Drevensek-Olenik, I.; Rupp, R.; Lenzini, G.; Lagerwall, J.P. High-fidelity spherical cholesteric liquid crystal Bragg reflectors generating unclonable patterns for secure authentication. *Sci. Rep.* **2016**, *6*, 26840. [CrossRef] [PubMed]
19. Kim, J.G.; Park, S.Y. Photonic Spring-Like Shell Templated from Cholesteric Liquid Crystal Prepared by Microfluidics. *Adv. Opt. Mater.* **2017**, *5*, 1700243. [CrossRef]
20. Seo, H.J.; Lee, S.S.; Noh, J.; Ka, J.W.; Won, J.C.; Park, C.; Kim, S.H.; Kim, Y.H. Robust photonic microparticles comprising cholesteric liquid crystals for anti-forgery materials. *J. Mater. Chem. C* **2017**, *5*, 7567–7573. [CrossRef]
21. Utada, A.S.; Lorenceau, E.; Link, D.R.; Kaplan, P.D.; Stone, H.A.; Weitz, D.A. Monodisperse Double Emulsions Generated from a Microcapillary Device. *Science* **2005**, *308*, 537. [CrossRef] [PubMed]
22. Faryad, M.; Lakhtakia, A. The circular Bragg phenomenon. *Adv. Opt. Photonics* **2014**, *6*, 225. [CrossRef]

23. St. John, W.D.; Fritz, W.J.; Lu, Z.J.; Yang, D.K. Bragg reflection from cholesteric liquid crystals. *Phys. Rev. E* **1995**, *51*, 1191–1198. [CrossRef]
24. Kim, S.H.; Kim, S.H.; Yang, S.M. Patterned Polymeric Domes with 3D and 2D Embedded Colloidal Crystals using Photocurable Emulsion Droplets. *Adv. Mater.* **2009**, *21*, 3771–3775. [CrossRef]
25. Noh, K.G.; Park, S.Y. Smart molecular-spring photonic droplets. *Mater. Horiz.* **2017**, *4*, 633–640. [CrossRef]
26. Tseng, H.F.; Cheng, M.H.; Li, J.W.; Chen, J.T. Solvent On-Film Annealing (SOFA): Morphological Evolution of Polymer Particles on Polymer Films via Solvent Vapor Annealing. *Macromolecules* **2017**, *50*, 5114–5121. [CrossRef]

nanomaterials

MDPI

Article

Templated Sphere Phase Liquid Crystals for Tunable Random Lasing

Ziping Chen [1], Dechun Hu [1], Xingwu Chen [2], Deren Zeng [2], Yungjui Lee [2], Xiaoxian Chen [2] and Jiangang Lu [1,*]

[1] National Engineering Lab for TFT-LCD Materials and Technologies, Department of Electronic Engineering, Shanghai Jiao Tong University, Shanghai 200240, China; positive_ping@sjtu.edu.cn (Z.C.); hdc86466240@126.com (D.H.)

[2] Shenzhen China Star Optoelectronics Technology Co., Ltd., Shenzhen 518132, China; chenxingwu01@tcl.com (X.C.); eden.tseng@tcl.com (D.Z.); kc.lee@tcl.com (Y.L.); hanks.chen@tcl.com (X.C.)

* Correspondence: lujg@sjtu.edu.cn; Tel.: +86-21-3420-7914

Received: 30 September 2017; Accepted: 7 November 2017; Published: 15 November 2017

Abstract: A sphere phase liquid crystal (SPLC) composed of three-dimensional twist structures with disclinations among them exists between isotropic phase and blue phase in a very narrow temperature range, about several degrees centigrade. A low concentration polymer template is applied to improve the thermal stability of SPLCs and broadens the temperature range to more than 448 K. By template processing, a wavelength tunable random lasing is demonstrated with dye doped SPLC. With different polymer concentrations, the reconstructed SPLC random lasing may achieve more than 40 nm wavelength continuous shifting by electric field modulation.

Keywords: sphere phase; lasers and laser optics; materials; liquid crystal

1. Introduction

A self-assembly sphere phase liquid crystal (SPLC)—consisting of three-dimensional twist sphere (3-DTS) structures and disclinations among them—exists in a narrow temperature range, approximately several degree centigrade, between isotropic phase and blue phase [1]. Due to the fast switching with low electric field, the SPLC attracts people's attention for its potential applications in displays, light shutters, and phase modulators after the temperature range is broadened to more than 358 K by stabilizing the disclinations with polymer networks [1]. By multiple scattering and interference effects in a chaotic amplifying medium, random lasing has appeared in scattering materials, such as polymer film [2], biological tissues [3], and liquid crystal [4]. Because of the 3-DTS structures, SPLC shows great potential application for random lasing. Recently, a SPLC random laser has been demonstrated with low threshold energy but weak thermal stability [5].

In this paper, a sphere phase template is demonstrated to improve the thermal stability of sphere phase. With LC refilling to the template, the temperature range of reconstructed sphere phase LC can be enlarged to more than 448 K. A random laser of reconstructed SPLC with wide temperature range is proposed. With the template of different polymer concentrations, a central wavelength tunable sphere phase random laser, whose tunable range is approximately 40 nm, can be achieved by electric field modulation. Therefore, templated SPLC shows great potential for photonic applications.

2. Materials and Methods

To investigate the reconstruction capability of the SPLC template, the material systems including 77.75 wt % of positive nematic LC (SP001, Δn = 0.148, $\Delta \varepsilon$ = 33.2, Jiangsu Hecheng Display Technology Co., Ltd., Jiangsu, China, (HCCH)), 4.21 wt % of chiral dopant (R5011,HCCH), 8.86 wt % of ultraviolet (UV)-curable monomer (12A,HCCH), 9.08 wt % of cross-linker agent (RM-257,HCCH), and 0.1 wt % of photo-initiator (IRG184,HCCH) were used in the experiment. The homogeneous mixture was capillary filled into the cell at the isotropic phase (353 K). The phase-transition process of the mixture was observed under a polarized optical microscope (POM, XPL-30TF, Shanghai WeiTu Optics & Electron Technology Co., Ltd., Shanghai, China) when it was cooled down from isotropic phase to chiral nematic phase at a rate of 0.5 °C/min by the temperature controller (HCS302, Intec Co., Ltd., Tokyo, Japan). The mixture showed the following phase sequence: isotropic phase -324.7 K-sphere phase -322.3 K-sphere phase and blue phase -321.5 K-blue phase- 308K-chiral nematic phase (N*). As shown in Figure 1b, sphere phase appeared between 324.7 K and 322.3 K with the light scattering phenomenon. From 322.3 K to 321.5 K, the coexistence of sphere phase and blue phase was observed with the light scattering and the classical platelet texture of the blue phase, as illustrated in Figure 1c.

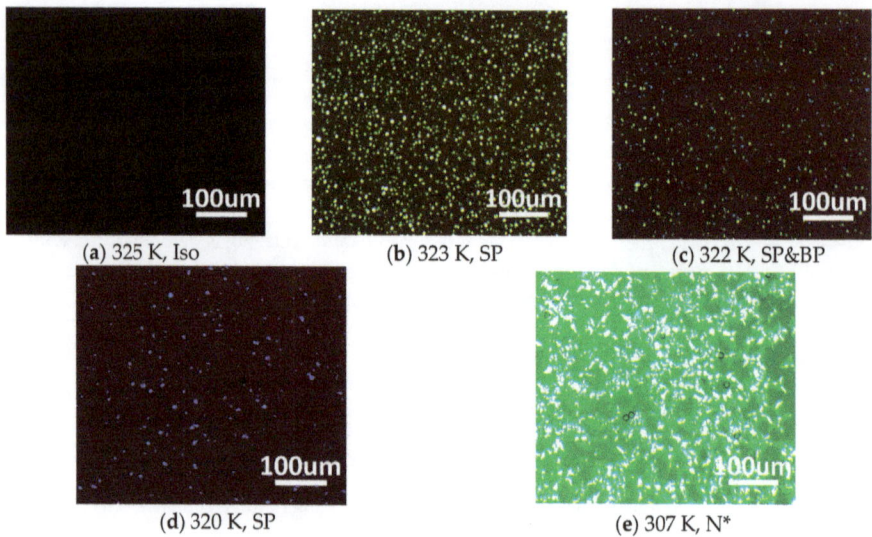

Figure 1. Reflective photographs under a POM of the LC mixture at different temperatures: (**a**) isotropic phase; (**b**) sphere phase; (**c**) sphere phase and blue phase; (**d**) blue phase; and (**e**) chiral nematic phase.

The sample was then cooled to 322.5 K in the temperature controller and irradiated with ultraviolet light (365 nm) at an intensity of 3 mW/cm^2 for 15 min and after polymerization, the transmission photograph of the polymer-stabilized sphere phase liquid crystal (PS-SPLC) under a POM at room temperature was shown in Figure 2a. The transition temperature from sphere phase to isotropic phase is 348 K. Then the cell was immersed in the acetone for about 48 h to wash-out the liquid crystal, chiral dopant, unreacted monomers, and photo-initiator. After evaporating the remaining acetone at 358 K, the free-standing porous sphere phase template was formed, as shown in Figure 2b. To confirm the reconstruction capability of the sphere phase template, the nematic LC, SP001, was refilled into the polymer templates at isotropic phase.

After LC refilling, the sphere phase texture was observed under a POM when the sample was cooled down to the room's temperature, approximately 298 K, as shown in Figure 2c. The result indicated that the achiral liquid crystal could be reconstructed to the SPLC due to the anchoring energy of the polymer template. The templated SPLC showed perfect thermal stability that its temperature range of sphere phase was approximately 447 K, from 173 K to 347 K, as illustrated in Figure 3.

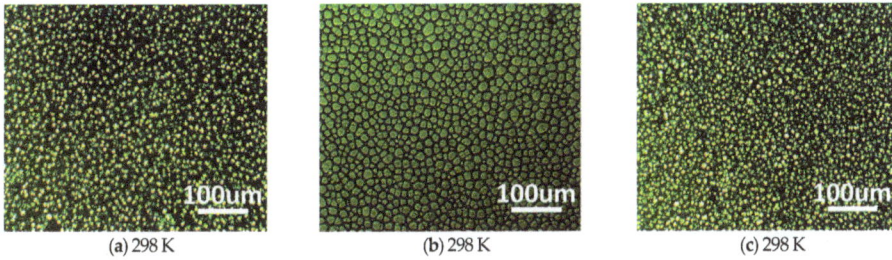

Figure 2. Transmission image of (**a**) the original PS-SPLC; (**b**) the polymer template; and (**c**) the templates with refilling nematic LC.

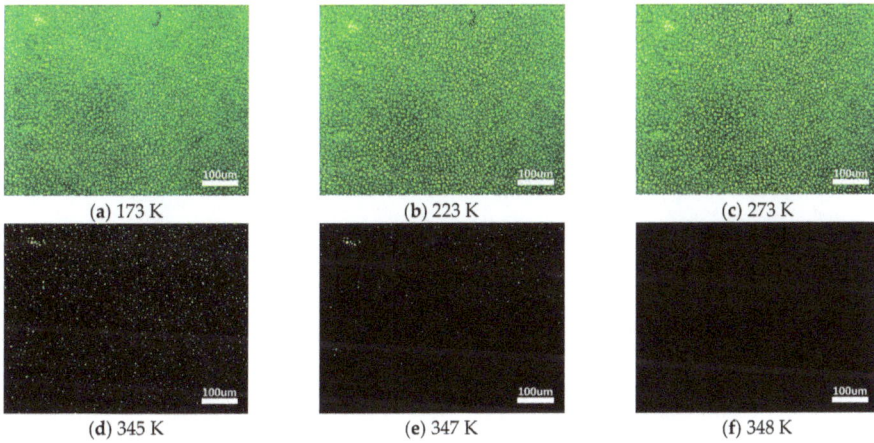

Figure 3. Transmission images of the templated SPLC phase transition from 173 K to 348 K.

Precursors with different polymer concentration, as listed in Table 1, were prepared to investigate the relationship between polymer concentration and the reconstruction capability of polymer template. As illustrated in Figure 4, only the cell of 14 wt % polymer concentration showed the texture of chiral nematic phase, four other cells of higher polymer concentration showed the sphere phase textures. According to our previous research [6], with the material systems listed in Table 1, 16 wt % polymer concentration was a threshold value for the SPLC reconstruction. The polymer template with polymer concentrations of 16, 18, 20, and 22 wt % provided enough anchoring energy to reassemble 3-DTS structure resulting in the SPLC reconstruction.

Figure 4. *Cont.*

(d) 20 wt %

(e) 22 wt %

Figure 4. Transmission images after refilling the nematic LC into the polymer template with the polymer concentration of (**a**) 14 wt %; (**b**) 16 wt %; (**c**) 18 wt %; (**d**) 20 wt %; and (**e**) 22 wt %, respectively.

Table 1. Precursors with different polymer concentrations.

Polymer Concentration	14 wt %	16 wt %	18 wt %	20 wt %	22 wt %
SP001 (wt %)	81.89	79.67	77.63	75.79	73.93
R5011 (wt %)	3.97	4.09	4.24	4.38	4.52
RM257 (wt %)	7.03	8.20	9.13	10.12	10.78
12A (wt %)	7.01	8.04	8.90	9.61	10.67
IRG184 (wt %)	0.1	0.1	0.1	0.1	0.1

3. Results and Discussion

To confirm a wide temperature wavelength tunable random lasing with the templated SPLC, a mixture of LC (99.7 wt %, SP001) and laser dye (0.3 wt %, Pyrromethene-597) is refilled into the cells with top-down ITO (Indium Tin Oxide) electrodes and 20 um cell gap of four kinds of SPLC templates with polymer concentrations of 16, 18, 20, and 22 wt % (Samples A, B, C, and D).

The experimental setup is shown in Figure 5. The cell is pumped by a Q-switched Nd:YAG (yttrium aluminium garnet) laser ($\lambda = 532$ nm; pulse width = 8 ns) with a repetition rate of 1 Hz. The pump beam is divided into two paths by a beam splitter. One is detected by an energy meter and the other is used as a pump source. The emission signals passed through the focusing lens are then collected by an optical fiber that is connected to a spectrometer (HR4000, Ocean Optics, Edinburgh, UK) [7].

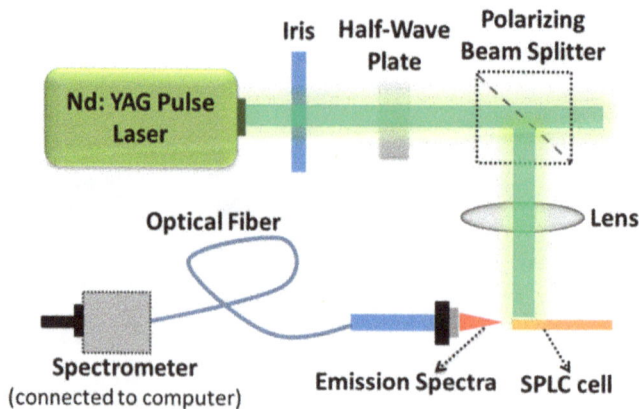

Figure 5. Experimental setup used to investigate laser action in dye-doped templated SPLC.

Rays emitted from the amplifying random medium travel among the self-assembled 3-DTSs of SPLCs while undergoing multiple scattering. When gain exceeds the loss, random lasing will occur [8–20]. As shown in Figure 6, laser emission can be observed in all the samples and the full width at half-maximum (FWHM) of the samples are approximately 6 nm. According to several prior studies of random lasing [21,22], the FWHM of the random laser was about 3 nm to 6 nm. Therefore, although the FWHM of the random laser is a little broad, we still relegate it to the random laser. We also think that the broad FWHM is probably the special characteristic of the random laser. As the polymer concentration of the templated SPLC system increases from 16 wt % to 22 wt %, the chiral dopant increases from 3.97 wt % to 4.52 wt %, resulting in decrease of the pitch length and blue shift of the central wavelength of the random lasing from 588 nm to 563 nm [20]. As listed in Table 1, the concentration of the chiral dopant rises from 4.09 wt % to 4.52 wt % when we make four kinds of the polymer templates, the helical pitch of the reconstructed SPLC decreases with the increase of polymer concentration, resulting in the shift of the central wavelength of the random lasing. As illustrated in Figure 7, the measured emission intensity of four samples is almost same because the LC molecules of four samples keep the same. Therefore, the wavelength tunable random lasing may be achieved with the templated SPLCs of different polymer concentrations. To further characterize the emission properties, the emitted intensity of sample B is recorded as a function of pump energy, as depicted in Figure 8. The lasing threshold is 4.01 nJ/pulse at room temperature.

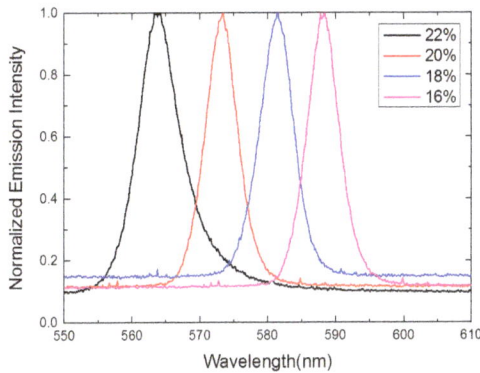

Figure 6. Emission spectra of templated SPLC systems with different concentrations of the polymer.

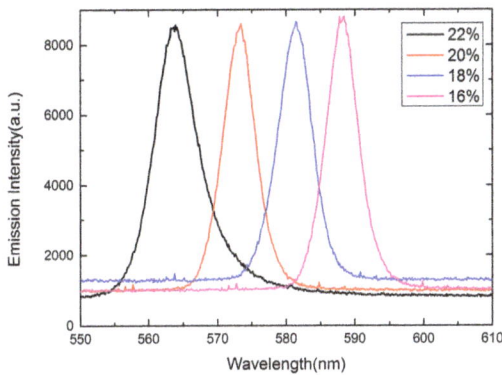

Figure 7. Measured emission spectra of templated SPLC systems with different concentrations of the polymer.

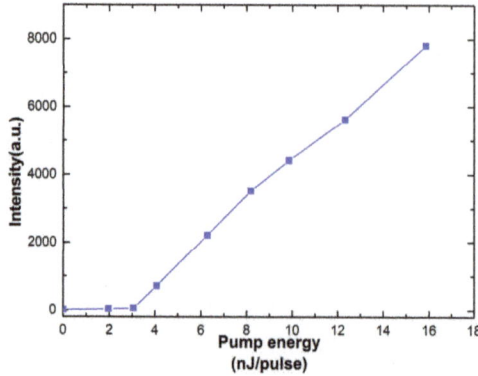

Figure 8. The emitted intensity of sample B as a function of pump energy.

Besides, to shift the wavelengths of the random lasing continuously, electric field tuning random lasing is proposed. All the samples are applied with a 1 kHz AC signal. As shown in Figure 9, when the electric field increases from 0 V/μm to 8 V/μm, the central wavelength of Samples A, B, C, and D shifts from 588 nm to 604 nm, 581 nm to 593 nm, 573 nm to 588 nm, and 563 nm to 579 nm, respectively. The central wavelength of all the samples shows red shift with increasing the electric field. If the vertical electric field increases 1 V/μm, the central wavelength of laser emission will generate about 2 nm red shift. Because the director of LC molecule is gradually changed with the increase of electric field, the cubic lattice of the SPLC is deformed, which shifts the central wavelengths of the templated SPLC random lasing. If the electric field is higher than 8 V/μm, the laser emission will disappear because the sphere phase change to the chiral nematic phase, resulting in the lack of random path among 3-DTSs. Figure 10 shows the measured emission spectra of sample D. As the electric field increased from 0 V/μm to 8 V/μm, the intensity of the random laser decreased. While the applied voltage increases, all LC molecules gradually reorient to the direction of the external field, resulting in gradual decrease of the multiple scattering. Therefore, with four kinds of the SPLC templates, a mixture of LC and laser dye, a continuous central wavelength shifting of the SPLC random lasing can be enlarged to 40 nm by the electric field modulation which shows great potential application in the commercial random laser. We used the relatively thick, 20 um, cell because we try to measure the lasing spectra in the three orthogonal directions, *x*, *y*, and *z*. The measured scattering intensity of the random lasing shown in Figure 11 is similar to the previous research [23]. So the beam divergence of the sphere phase random laser is in the medium level.

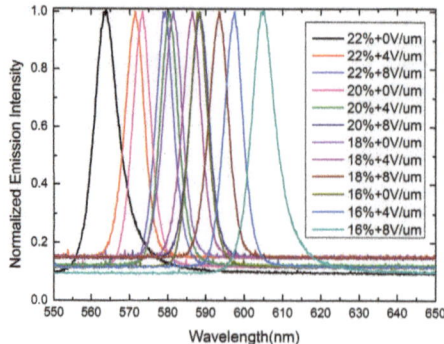

Figure 9. Emission spectra of templated SPLC systems under different electric fields.

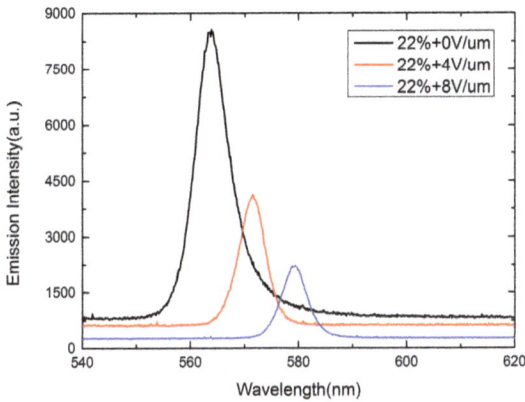

Figure 10. Measured emission spectra of sample D under different electric fields.

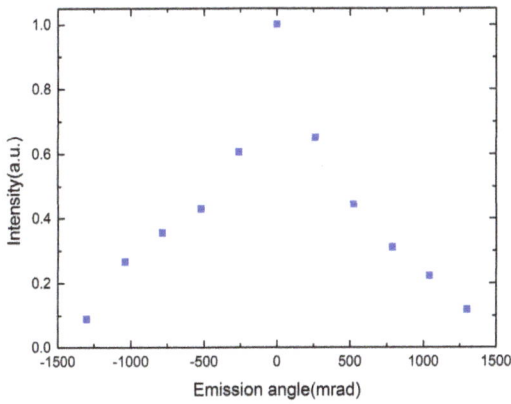

Figure 11. The measured scattering intensity of random lasing.

4. Conclusions

In summary, we propose a sphere-phase template to improve the temperature range of sphere phase LC to more than 448 K. Based on the templated SPLC, a central wavelength tunable sphere phase random lasing is demonstrated. With different concentrations of chiral dopants, the random lasing from the reconstructed sphere phase may achieve a more than 40 nm wavelength shifting by electric field modulation.

Acknowledgments: This work was sponsored by National High Technology Research and Development Program of China (2015AA017001), NSFC (61775135), and financially supported by Shenzhen China Star Optoelectronics Technology Co., Ltd.

Author Contributions: Jiangang Lu conceived and designed the experiments; Ziping Chen, Dechun Hu, Xingwu Chen, Deren Zeng, Yungjui Lee, and Xiaoxian Chen performed the experiments; Ziping Chen analyzed the data; Ziping Chen and Jiangang Lu wrote the paper.

Conflicts of Interest: The authors declare no conflict of interest.

References

1. Zhu, J.L.; Ni, S.B.; Chen, C.P.; Wu, D.Q.; Song, X.L.; Chen, C.Y.; Lu, Y.K.; Shieh, H.D. Chiral-induced self-assembly sphere phase liquid crystal with fast switching time. *Appl. Phys. Lett.* **2014**, *104*, 091116. [CrossRef]
2. Frolov, S.V.; Vardeny, Z.V.; Yoshino, K.; Zakhidov, A.; Baughman, R.H. Stimulated emission in high-gain organic media. *Phys. Rev. B* **1999**, *59*, R5284–R5287. [CrossRef]
3. Polson, R.C.; Vardeny, Z.V. Random lasing in human tissues. *Appl. Phys. Lett.* **2004**, *85*, 1289–1291. [CrossRef]
4. Wiersma, D.S.; Van, M.P.; Lagendijk, A. Coherent backscattering of light from amplifying random media. *Appl. Phys. Lett.* **2015**, *75*, 1739–1742. [CrossRef] [PubMed]
5. Zhu, J.L.; Li, W.H.; Sun, Y.B.; Lu, J.G.; Song, X.L.; Chen, C.Y.; Zhang, Z.D.; Su, Y.K. Random laser emission in a sphere-phase liquid crystal. *Appl. Phys. Lett.* **2015**, *106*, 191903. [CrossRef]
6. Hu, D.C.; Li, W.H.; Chen, X.W.; Ma, X.L.; Lee, Y.J.; Lu, J.G. Template effect on reconstruction of blue phase liquid crystal. *J. Soc. Inf. Disp.* **2016**, *24*, 593–599. [CrossRef]
7. Zhou, Y.; Huang, Y.H.; Ge, Z.B.; Chen, L.P.; Hong, Q.; Wu, T.X.; Wu, S.T. Enhanced photonic band edge laser emission in a cholesteric liquid crystal resonator. *Phys. Rev. E* **2006**, *74*, 061705. [CrossRef] [PubMed]
8. Wiersma, D.S. The physics and applications of random lasers. *Nat. Phys.* **2008**, *4*, 359–367. [CrossRef]
9. Morris, S.M.; Ford, A.D.; Pivnenko, M.N.; Coles, H.J. Enhanced emission from liquid-crystal lasers. *J. Appl. Phys.* **2005**, *97*, 023103. [CrossRef]
10. Furnmi, S.; Yokoyama, S.; Otomo, A.; Mashiko, S. Electrical control of the structure and lasing in chiral photonic band-gap liquid crystals. *Appl. Phys. Lett.* **2003**, *82*, 16–18. [CrossRef]
11. Kasano, M.; Ozaki, M.; Yoshino, K.; Ganzke, D.; Haase, W. Electrically tunable waveguide laser base on ferroelectric liquid crystal. *Appl. Phys. Lett.* **2003**, *82*, 4026–4028. [CrossRef]
12. Lin, T.H.; Jau, H.C.; Chen, C.H.; Chen, Y.J.; Wei, T.H.; Chen, C.W.; Fuh, A.Y. Electrically controllable laser based on cholesteric liquid crystal with negative dielectric anisotropy. *Appl. Phys. Lett.* **2006**, *88*, 061122. [CrossRef]
13. Chanishvili, A.; Chilaya, G.; Petriashvili, G.; Barberi, R.; Bartolino, R.; Cipparrone, G.; Mazzulla, A.; Oriol, L. Phototunable lasing in dye-doped cholesteric liquid crystals. *Appl. Phys. Lett.* **2003**, *83*, 5353–5355. [CrossRef]
14. Furumi, S.; Yokoyama, S.; Otomo, A.; Mashiko, S. Phototunable photonic bandgap in a chiral liquid crystal laser device. *Appl. Phys. Lett.* **2004**, *84*, 2358–2363. [CrossRef]
15. Shibaev, P.V.; Sanford, R.L.; Chiappetta, D.; Milner, V.; Genack, A.; Bobrovsky, A. Light controllable tuning and switching of lasing in chiral liquid crystals. *Opt. Express* **2005**, *13*, 2358–2363. [CrossRef] [PubMed]
16. Ohta, T.; Song, M.H.; Tsunoda, Y.; Nagata, T.; Shin, K.C.; Araoka, F.; Takanishi, Y.C.; Ishkawa, K.; Watanabe, J.J.; Nishimura, S.; et al. Monodomain film formation and lasing in dye-doped polymer cholesteric liquid crystals. *Jpn. J. Appl. Phys.* **2004**, *43*, 6142. [CrossRef]
17. Shibaev, P.V.; Kopp, V.; Genack, A.; Hanelt, E. Lasing from chiral photonic band gap materials based on cholesteric glasses. *Liq. Cryst.* **2003**, *30*, 1391–1400. [CrossRef]
18. Finkelmann, H.; Kin, S.T.; Munoz, A.; Muhoray, P.P.; Taheri, B. Tunable mirrorless lasing in cholesteric liquid crystalline elastomers. *Adv. Mater.* **2001**, *13*, 1069–1072. [CrossRef]
19. Morris, S.M.; Ford, A.D.; Pivnenko, M.N.; Coles, H.J. Electronic control of nonresonant random lasing from a dye-doped smectic A* liquid crystal scattering device. *Appl. Phys. Lett.* **2005**, *86*, 141103. [CrossRef]
20. Huang, Y.H.; Zhou, Y.; Doyle, C.; Wu, S.T. Tuning the photonic band gap in cholesteric liquid crystals by temperature-dependent dopant solubility. *Opt. Express* **2006**, *14*, 1236–1242. [CrossRef] [PubMed]
21. Cao, H.; Zhao, Y.G.; Ong, H.C.; Chang, R.P.H. Far-field characteristics of random lasers. *Phys. Rev. B* **1999**, *59*, 15107–15111. [CrossRef]
22. Woltman, S.J.; Crawford, G.P. Tunable cholesteric liquid crystals lasers through in-plane switching. *Proc. SPIE* **2007**, *6487*. [CrossRef]
23. Cao, H.; Zhao, Y.G.; Ho, S.T.; Seeling, E.W.; Wang, Q.H.; Chang, R.P.H. Random laser action in semiconductor power. *Phys. Rev. Lett.* **1999**, *82*, 2278–2281. [CrossRef]

MDPI

St. Alban-Anlage 66

4052 Basel

Switzerland

Tel. +41 61 683 77 34

Fax +41 61 302 89 18

www.mdpi.com

Nanomaterials Editorial Office

E-mail: nanomaterials@mdpi.com

www.mdpi.com/journal/nanomaterials

www.ingramcontent.com/pod-product-compliance
Lightning Source LLC
Chambersburg PA
CBHW051904210326
41597CB00033B/6022